Listening, Belonging, and Memory

Listening, Belonging, and Memory

Abigail Gardner

BLOOMSBURY ACADEMIC
NEW YORK • LONDON • OXFORD • NEW DELHI • SYDNEY

BLOOMSBURY ACADEMIC
Bloomsbury Publishing Inc
1385 Broadway, New York, NY 10018, USA
50 Bedford Square, London, WC1B 3DP, UK
29 Earlsfort Terrace, Dublin 2, Ireland

BLOOMSBURY, BLOOMSBURY ACADEMIC and the Diana logo are trademarks of Bloomsbury Publishing Plc

First published in the United States of America 2023
This paperback edition published 2025

Copyright © Abigail Gardner, 2023

For legal purposes the Acknowledgments on p. xi constitute an extension of this copyright page.

Cover design: Louise Dugdale
Cover image: Sally Nash and the Del Turner Quintet © Veterans' Voices
http://veteransvoicesglos.co.uk

All rights reserved. No part of this publication may be reproduced or transmitted in any form or by any means, electronic or mechanical, including photocopying, recording, or any information storage or retrieval system, without prior permission in writing from the publishers.

Bloomsbury Publishing Inc does not have any control over, or responsibility for, any third-party websites referred to or in this book. All internet addresses given in this book were correct at the time of going to press. The author and publisher regret any inconvenience caused if addresses have changed or sites have ceased to exist, but can accept no responsibility for any such changes.

Library of Congress Cataloguing-in-Publication Data
Names: Gardner, Abigail, author.
Title: Listening, belonging, and memory / Abigail Gardner.
Description: New York : Bloomsbury Academic, 2023. | Includes bibliographical references and index. | Summary: "An interdisciplinary approach to listening, in relation to the self, nation, age, witnessing and memory"-- Provided by publisher.
Identifiers: LCCN 2022059006 (print) | LCCN 2022059007 (ebook) | ISBN 9781501376801 (hardback) | ISBN 9781501376849 (paperback) | ISBN 9781501376818 (ebook) | ISBN 9781501376825 (pdf) | ISBN 9781501376832 (ebook other)
Subjects: LCSH: Listening. | Belonging (Social psychology)
Classification: LCC BF323.L5 G344 2023 (print) | LCC BF323.L5 (ebook) | DDC 153.6/8--dc23/eng/20230302
LC record available at https://lccn.loc.gov/2022059006
LC ebook record available at https://lccn.loc.gov/2022059007

ISBN: HB: 978-1-5013-7680-1
PB: 978-1-5013-7684-9
ePDF: 978-1-5013-7682-5
eBook: 978-1-5013-7681-8

Typeset by Deanta Global Publishing Services, Chennai, India

To find out more about our authors and books visit www.bloomsbury.com and sign up for our newsletters.

Contents

List of Figures — vi
Preface — vii
Acknowledgments — xi

Introduction: Life as a Listener — 1
1 Connecting Lineages — 15
2 Applying Connected Listening — 39
3 Listening Across Age(s) — 63
4 Listening and Belonging — 91
5 Listening: Migration, Voice, and Place — 121
6 Echoes — 151

Bibliography — 155
Index — 170

Figures

2.1	Veterans' Voices Project	52
3.1	Page 3 of Alan's story	67
3.2	Sally (right) and her dancers	69
3.3	Sally Nash and the Del Turner Quintet	70
3.4	Genevieve Levy	83
4.1	C's Stained-glass Stag	109
5.1	MaMuMi Song Story Questions	129
6.1	Robert Landberger's boat ticket from Holland to England, 1938	153

Preface

Listen. Who Do You Hear?

This is a book about listening to voices. Sometimes these voices tell stories about themselves, or about the music they like, sometimes these voices come across the radio, or are archived in national sound collections. Sometimes the voices tell stories about their pasts, sometimes images tell stories about the past because the voices are gone. Sometimes you have to strain to hear the voices through swathes of silence and story. And it is these two, silence and story, that lie at the heart of the book, how they interweave, how they are activated, maneuvered, reconfigured, and denied. This is a process that is also about foregrounding the ordinary, listening for resonance and experience from spaces and people whose narratives have slipped beneath accepted historiographies, expectations, and statistics. It is about listening in action, and observations from the "field" of digital storytelling. And it is about listening online, a practice which extends to social media threads, newspaper articles, and UK government policy statements. Using these digital storytelling projects, social media, broadcast history, and current affairs as examples, it argues that listening is done with age, and as witness, in place, and in time. With reflections on how listening may be changed by where we listen, when and who to, it argues for the importance of understanding the crucial role that listening has in contemporary media and sound cultures.

    A contemporary snapshot of applied media and popular music research and critical media analysis, it is written from a cultural studies perspective, which prioritizes the importance of lived experience, pleasure, and expression, and is an attempt to place a specific kind of listening at the center of current debates about which voices might be listened to, who by, and why. It is avowedly political in its bid to place "connected listening" center stage as a micro mode which can counteract authority and its stories, and its methods draw from feminist politics of care to open up enunciative spaces to afford witnessing and agency (LaBelle 2018; Western 2020; Berlant 2011; Stewart 2007).

I am using listening as a mode of giving attention to someone, or to something. I take this idea of "giving attention" to argue that this process is complicated by who, where, and when we listen. There are many connections to unravel when listening, many lines back through, across, forwards. Listening is about being connected to ourselves, our surroundings, to sound, to songs, to others. It is a moment, an action, within a knotted network (Ingold 2015), woven into multiple threads of perception. What follows here are some snippets, a montage of some listening episodes to start with, that speak to the areas I am interested in, listening across age(s), being listened to and not being listened to, listening and memory, and belonging.

Listening to the Passed

After my father died in 2004, my mother was unable to change the answer message on their phone. Every time I called her, I heard him. A soft Somerset burr to the voice, acquired over forty years of living there, a voice that told me he was unable to get to the phone. It took my mother months to delete that message. She never did get rid of any of his clothes; his side of the wardrobe remained the same as before the summer of 2004. But it was his voice, coming to me in the present, from a body that had died, which both cut me up and comforted me. Where was my father when I was listening to him? He was both not there and there, a sonic trace of his body. He was gone but those sound vibrations were recorded onto that home phone tape. Something of him still resonated in the world. And at some point, it became too painful to keep listening and she deleted it.

Listening to the Past

Every morning at 10.47 am GMT, the ice-cream van came round. It was January, February 2021. It was cold. Covid-19 was stalking. No schools were open. The same jingle played; jolly, baroque, muffled in that way the ice-cream jingles are. I had conversations with friends about whether or not it was selling ice-cream; perhaps it was a drugs front? And every day at 10.50 it took me back to the cul-de-sac in Somerset where I grew up, and from there to beaches and parks and school gates, to queuing up for a 99 with my own children. Listening is a portal to our pasts, sometimes one we are ushered through involuntarily. I never saw the van.

The Listening State

In April 2021 the Conservative government in the UK proposed a new Police, Crime and Sentencing Bill, to include the following:

> This measure will broaden the range of circumstances in which the police can impose conditions on protests, including a single person protest, to include where noise may cause a significant impact on those in the vicinity or serious disruption to the running of an organisation.[1]

In the middle of May 2021, Israel saw its worst violence for years. There were escalating confrontations between Israelis, Israeli Arabs, and Palestinians. A UK newspaper, *The Guardian*, ran a headline about Gaza which declared how "Israel vows not to stop attacks until there is 'complete quiet.'"[2]

This is the politics of authority using the language of sound. These state listeners do not want to hear dissent, to have their sonic (and political) environment disturbed by noise. The UK government holds the power over who else may "sound out", and who has to keep quiet. Noise as protest, noise as disturbance is decried and silence is set as the standard. These two governments were operating a mode of control via sonic policing.

These instances show how listening is mobile. It is a fulcrumic activity, qualified by where it takes place and who is doing it. It is part of a network of power relations, intimate and institutional. It is the portal to pasts, and to scraps of memory, reassembled in the light of an auditory impulse. It is part of a circuit of communication happening at unexpected "junctions" (Deleuze and Guattari 1994: 23 in Schulte 2016: 147).

Edge Work

During the zoom time of Covid-19, I attended a session on Deep Listening run by Ximena Alarcón Díaz, a sound artist and researcher into relational listening.[3] Called "Voice and Listening: Techniques for Political Life" held on March 25, 2021, "at" Birmingham City University. I was asked to stand up in front of my

---

[1] https://www.gov.uk/government/publications/police-crime-sentencing-and-courts-bill-2021-factsheets/police-crime-sentencing-and-courts-bill-2021-protest-powers-factsheet
[2] https://www.theguardian.com/world/2021/may/12/israel-only-stop-gaza-attacks-when-complete-quiet
[3] https://www.ximenaalarcon.net/

laptop (itself a radical demand at the time), to shake, feel my feet grounded in the earth, shut my eyes and to listen. At first, I was reluctant, having a fear of all things "group" like. But then I listened. For five minutes, myself and a selection of academics, practitioners, artists did the same. I heard buses idling at the junction outside my house, people talking, and footsteps. I was asked to share my experiences in a breakout room. Another participant talked about listening to her fridge. I was unable to disentangle my listening of the bus, to a visual image of it outside, to a real dislike for the vehicle and its idling diesel fumes coming down to my basement. I listened to the rhythm of its engine, expecting it to start up and pull away from the junction. This was expectant listening, listening out for something.

It was an enjoyable session, but this book is not about Deep Listening, nor is it about listening to the beyond human or the natural world. This is a theoretical world I am becoming aware of but cannot, as yet write about. I respect the listening work that is being done by many sound studies scholars (Donovan 2019; Ouzounian 2021; Schulze 2020) which has opened my ears to how many "listenings" there are: acousmatic, causal, emergent, expanded, every day, immanent, multiplicity of, musical, ordinary and specialist, reduced, reflective, re-imagining of, semantic, process, physiological, psychological, verb, concept, entanglement, incursion, reward, and punishment. The listening that I am doing here is built around narrative, story, image and voice and their interrelationships and is a reflection on applied research that I have done "on the edges" of previous work on women, aging, and popular music (Jennings and Gardner 2012, Gardner 2020; Gardner and Jennings 2020). Coming from cultural studies and popular music studies, I have listened to people whose voices lie at the edges of the heard. I want to share the sound of their voices, for a while, for a moment.

Acknowledgments

Thanks to all of those who have let me listen, and who have listened with me. Chris Blick, Tommy Clough, Alan Dowdeswell, Emlyn Pennah, Rick Rutter, Sally Welzel, Brian Woolaston, Helen Atkinson, Chris Walker, Abil Gunawan, Jack Curtis, Angeline Xuan, Jack Reese, Lewis Ingham, Louise de Freitas, Megan Smith, Phoebe Nott, Sophie Flowers, Julia Hayball, Gideon Capie, Adam Behr, Annita Tsolaki, Afroditi Azari, Dario di Ferrante, Daniel Pittl, Lora Yoncheva, Kai Arne Hansen, Camilla Kvaal, Annabelle Holder, Anna Newell, Zayd Dawood, Thiago Viana, Nick Ives, Toby Harris-St John, and the twenty-nine people who shared their song stories with us. Paul Courtney, Leonie Burton, Fahimeh Malekinezhad, Jack Higgins, Ivor Richards, the GEM participants, Donna, Janet Trotter, Nigel Hatten, Louise Livesey, Abi Fisher, Jenny Silverston, Alan Silverston, Léa Elliott, Robert Landberger, Paul Zinder, Pawel Sawicki, Faye Hatcher. Damian Maye, Julie Ingram, Pippa Simmonds, and Sophia Raseta. Thanks to the wider IASPM network, including but not limited to, Sara Cohen, Sarah Hill, Richard Elliott, Line Grenier, Helmi Järviluoma-Mäkelä, Martin Cloonan, Tom Attah, Paul Long, Alex de Lacey, and Eric Smialek. Thanks to Anne Dawson, Tracy Symonds, Maria Quinn, Ros Jennings, Joanne Garde-Hansen, Tom Soper, Ben Wardle, Philip Reeder, Arran Stibbe, Jon Hobson, Kenny Lynch, Emily Ryall, Julia Peck, and the FCH librarians. Thanks to my family, Charles, Josh, Letty, and Louis, and my great friend, Anne. The book is for Sheila, whose listening made a difference.

Introduction

Life as a Listener

We all become migrant people by virtue of being listeners or readers. For truly regarded, all reading or listening—like all writing—is a kind of travelling.

Fraser 2018: 29

Listening is travel
Listening is exposure
Listening is proximity
Listening is collapse
Listening is disciplined
Listening is commemoration

This label is faded. It is written in pencil. You can just about make out the first name, just about. The other label is bold. No. 158. Two holes are punched into the top, on the left and on the right. A string holds them together. This label was to hang around a head. The faded label was secured to a suitcase. They belonged to Robert. He is gone now. But on April 6, 2022, his voice came out of a university lecture room's loudspeaker as part of an inaugural lecture. I was giving that lecture and was tracing back my interest in listening to 2009, when I worked on an oral history film commissioned by the Cheltenham Hebrew Congregation. Together with my then colleague Joanne Garde-Hansen, we were listening to elderly members of the community talk about their arrival in Cheltenham, where I lived, and where a 185-year-old synagogue is still home to a dwindling Orthodox community. Robert had not wanted to appear on camera, he thought he wasn't handsome enough, and so we used digital storytelling for him to tell his story as we did not need to film him, instead, we needed to take photographs of some of the objects he had from 1938 and to record his voice. He had the labels with which he had been identified when his mother put him on the Kindertransport train out of Vienna, via the Hook of Holland, to Harwich, where he was lodged in a holiday camp until a foster family came to house him.

He was one of the last to be picked because, as he says, "I wasn't blonde, and I didn't have curly hair." Robert's voice is deep and resonant. It was thirteen years since we had recorded his story. This clip, his voice, his humor was there in the room. It was felt. It was an act of sonic co-presence, bringing him into a listening relationship with the audience who were seated there in April 2022. This voice filled the room. It was a "sonic phenomena" (Western 2020: 304, see also James 2019), rolling around the lecture room's walls, into the ears of the audience, vibrating, landing. It was more than a "grain" (Barthes 1977), it had texture and weight. It was a sonic thing beyond meaning or signification and outside of that economy of meaning; it was affective (Ahmed 2004; Steward 2007) and it was also located, was placed somewhere. That is, while it was present, it was also tethered to a place in time and in history. It belonged somewhere(s). Where to? It was accented, the r's of "Kindertransport" trilled, the Austrian German acting out the bassline of the vocal beat. Robert never talked in his native language. Ever. It was there, beneath the English words, haunting the listening of him. Robert's voice was something of "value" (Couldry 2010 in Chouliaraki and Georgiou 2022: 110); it was claiming its place, bearing witness, and succeeding in being heard. Robert was still there. For those who didn't know him, he might never have gone.

This anecdote serves to set out how this book is about listening across bodies, and across time. It is about listening to voices and images, and to people talking about music. It is about listening to the past as it appears in the present. It is about the interplay between telling and listening and the narratives that emerge from those encounters. It is about listening in, and out of, spatial and temporal place. It is about listening as an organizing imperative, as a political act: part of a circuit of "speaking and acting together" which constitutes the "political realm" (LaBelle 2018: 9 citing Arendt 1998). It is about trans-listening, listening across disciplines, bodies, times.

Deborah Moglen (2008) has the lovely idea of "trans-aging," which is about time and experience. Her work is a psychoanalytic exploration of aging and the self; an attempt to think outside of the expected narrative trajectory, and it is something that comes up in the listening projects that I talk about in Chapter 3, where age and reflection confuse linearity. She argues for two models of thinking about aging, the vertical and the horizontal. The vertical model is driven by repression, whereas the horizontal is constituted by dissociation. This in turn takes two forms, the incorporative or the introjective function. The incorporative model is defensive, maintaining "ghostly specters of youth as consuming objects of loss and desire" (2008: 297). The introjective model, by contrast, "initiates a dynamic and creative process in which multiple self-states of past and present are available for recognition and enactment" (2008: 297). This embraces the "endlessly overlapping states of being and stages of life" (2008: 306). It starts to see the potential for a non-linear model of aging, which is disarticulated from a rigid chronology and characterized instead by relations, moments, experience, and memory. This model allows for an "overlapping," and, I think, an immediacy which I explore later in the book. It goes some way to articulating the experience of listening to Robert's voice as it filled the room and went into the ears of those present there on April 6, 2022; it coursed through live bodies, through senses (Barthes 1985a). I was taken back to the moment I first heard Robert and his voice prefigured a temporal collapse.

Listening, Belonging, and Memory

We listen in corporeal, familial, institutional, and ideological environments whose textures ebb and flow, whose modes and nuisances we negotiate. I listen for a child to come home at 3 a.m., to the emails popping into the inbox, to the ever-present I-Phone, I try to block out the sounds of my country's politicians vying for my attention, I listen out for oppositional support. There is a large literature on sound and our place within it, some of which I note in the next chapter, but the rationale for my work emerges, like most, from a view that there is a gap in the field of knowledge on the approach I take, and from an assessment of it as being core to much of the contemporary conceptual and practical work that I have conducted and am in the process of working on. There is no book that deals with "Listening" as a core methodology for use across media and popular music studies and its connections with belonging and memory, for using listening as a step into storytelling research. Les Back's brilliant book *The Art of Listening* (2007) did much to establish it as key to a sociological methodology, however, to use it as a legitimate way in to approaching creative music practices, oral history projects, and storytelling projects is to understand its importance and potential for listening as a cross-disciplinary methodology when conceived of as a process that happens within broader matrices of belonging and memory. Conceptually, the book is about listening as a trans-temporal and multi-sensorial mode, and these modes are worked out through examples of research projects that illustrate different circuits of listening.

The book is written from a cultural studies approach where popular culture and popular music act as valid locations from which to build the core arguments around listening as a multi-sensorial and temporal mode. I am not a philosophy scholar, nor am I a sound studies one. I work in the overlaps between popular music and media, oftentimes too "music" for media and too "media" for music. The research I have done drives this book, and so it might be argued that it is grounded in those voices and stories. This approach is founded on an understanding that the music, stories, narratives, and online media that serve as examples in the book are all doing some kind of work, that they are meaningful in and for themselves, that in becoming audible they might alter discourses of exclusion or of ignorance.

I think that I want the book to be able to work on two levels. First, I want to make it clear that listening is not the same as reception, and that understanding its many dimensions and roles reveals its increasing importance within online communication and community work. This means I am reliant on Barthes' idea from *The Responsibility of Forms*, where he writes about how listening "speaks" (1985a: 259). He fuses the auditory with the enunciative, foregrounding the interaction between the two, that space of exchange here. This is a space that I look at in this book, paying attention to how we might open up such spaces within the context of applied media and music research. Second, I want to use "Listening" in order to travel across disciplines, and so I use a variety of cross-disciplinary avenues into thinking about being a listener, life as a listener, listening to, listening for, in chapters that encompass reflections on time, space, memory, and listening to accounts of projects in community media and music, on oral histories and storytelling.

Listening and Belonging

Being listened to, being rendered audible, is key to belonging. Given that listening happens in and between bodies, real, virtual, and imagined (fleshly, digital, and national/diasporic), I touch on the relationships between the two and note how the act of listening can be a weapon of exclusion as well as a tool for inclusion. In 2004, Sara Ahmed clarified her position on the idea of comfort, which she said was "about the fit between body and object" (Ahmed 2004: 148). She was reflecting on how heteronormative society is like an uncomfortable chair for the queer subject, who can never "fit." Similar in practical terms to the design of cars, whose driving seats are engineered for male drivers, Ahmed's work spotlights the heteronormative (and patriarchal) architecture of social and material life through and upon which those who are "different" need to travel and negotiate. Rather than a chair, let's think about a mold, or a piece of foam, which has form but can also be changed through use. Similar I guess, in spirit to Butler's (1990) concept of performativity, wherein gender becomes through repetition, belonging is molded through many series of encounters, refusals, and acceptances, and I explore this in Chapter 4 in relation to citizenship and the dynamics of inclusion secreted into specific processes of listening. Rather than bell hooks' idea of belonging as being about "a" place and community (2009), the book thinks about belonging in relation to time and memory.

Listening and Memory

Memory studies is a well-established and flourishing area of academic research, and my job here is to think about how it is related to listening. I am meeting memory studies through digital storytelling, and so my entry point is story. I acknowledge work on memory as an iterative process (Barad 2010; Hristova, Ferrandiz, and Vollmeyer 2020; Kuhn 2007, 2010), particularly in relation to image (Campt 2017; Kuhn 2007, 2010) as well as noting the role of popular music and memory as a trigger to tell stories (Istvandity 2019; Van Dijck 2014). Throughout this, I am holding on to "story," that is the desire for narrative, and the truths they might enunciate might be different to "historical" or "factual" truth (Spence 1982, in Hirsch and Spitzer 2009: 160), but is key to "listening across age(s)," as the vehicle through which memory is relayed to the self and encountered by others in listening encounters. Some of these encounters might be termed "witnessing events" and the idea of the witness comes out of memory studies in relation to the Holocaust. In particular, Felman and Laub's (1992) work which centers on psychoanalysis and the Holocaust is instrumental to this and there are two strands of their thinking that are useful: the idea that witnessing is a moment of subjectivity collapse, and that witnessing might be a productive act. They argue that in witnessing trauma "the listener to trauma comes to be a participant and a co-owner of the traumatic event: through his very listening, he comes to partially experience trauma in himself" (1992: 57) and this proximity, the collapse into the other is an act of exposure. The idea that listening might be a productive process is outlined here, where they write how "The listener [. . .]

is a party to the creation of knowledge *de novo*" (1992: 57), a place, a moment where "things flash up" (Stewart 2007: 68), and lines meet. Listening can thus be transformative, modulating relationships between the listener and the listened to. Listening brings us closer.

Listening and Proximity

To listen to something or someone is to be close to it or them. Listening is about proximity, and in many works on sound and listening, which are outlined in Chapter 1, it has been considered the opposite of the visual experience, which effects distance. Even at its most utilitarian, when it is about access and information, it is also about getting closer to that which is being listened to. I will examine this throughout the book by considering the politics of proximity in relation to enablement and containment. I say "proximity" to mean that listening as a methodology is about being enfolded into difference, which may also involve other narratives (Macarthur 2016: 7). This sense of being "enfolded into" is an important one, which has been considered in relation to the nascent individualized listening technologies of the late twentieth century. In his assessment of the emotional and cognitive affordances of the Walkman, Bull (2004) argues that such media technologies offer a sense of proximity, or a "we-ness" and "being with" (2004: 186). They also manufacture distance from the "real" environment by inserting a selected one (work on headphones). Depending on what is being listened to, an album, a playlist, an audio book, the radio, mobile technologies offer a temporary community with other voices, sounds, and stories. My work centers on those stories, rather than the technological mediums through which they are relayed, although I do consider the method of representation that Twitter feeds have in relation to listening in Chapter 3.

Close listening can also be uncomfortable or undesired, messy, and political (Oliver 2015). Music might be curtailed, voices silenced, "noisy" protests made illegal. Sound and music can be and have been used as a weapon, as torture (Cusick 2017; De Nora 2000). Although I do not consider that in this book, I do think about listening when there is a discomfort, a distaste, or even a bodily refusal to be in the same space, to be trapped in proximity, which I consider briefly in Chapter 4 in relation to sounds of Otherness that are dampened down, muted, and refused and ask why that might be.

Connected Listening

"Connected listening" is listening that happens within a matrix of real, virtual, collective, national, and diasporic bodies, which themselves are in constant flux. The listener listens from their body, and this exists within a web of identities, histories, and memories. A theory of connected listening weaves together ideas from sound studies (Kheshti 2014; Voegelin 2014; Thompson 2017; LaBelle 2018), cultural geography (Western 2020), feminist cultural theories of storytelling (Fernandes 2017), and affect (Ahmed 2004; Stewart 2007; Berlant 2011) to understand the role of listening as imbricated within complex and fluid fields of affective belongings that intersect across space and time. More specifically, the body is a thing that exists temporally and spatially; it is located. To think about how we might listen and how listening bodies might without thinking about where they might be found is to ignore the contexts within which that listening is done.

Listening happens within and between things and people, across bodies, memories, and imaginations. It has, as LaBelle writes, "relational affordances" (2018: 26). It travels, and in turn, enables travel. If so, we might ask where we are before we set off on our listening. In his 1993 book, *Weltfremdheit* (*Alienation*), German philosopher Sloterdijk asks, "Where are we when we listen to music"? ("Wo sind wir, wenn wir Musik hören?" (2016: 294). Now, this "where" might be in a spatial or a temporal "place" and his question is, I think, interested not only in the ontology of the listening subject but in the relational impact that listening has on it. I am not just thinking about listening to music, as I said earlier, but to voices. And so, my response to his question, "where are we . . .?," is that we are exposed and that this can be a positive place to be. We are open to invited and unintended sounds and stories. I am interested in the experience of listening encounters, between people, to music and to images, to the dynamics of listening between people and things. Sloterdijk's question, which focuses on one side of that dynamic, the listener, uses "wo" (where), which can be understood to encompass a chronological or locational "where." But let's replace that with "when," which upsets the security of the listening subject, forcing it to be reconsidered in relation to the encounter within which they are engaged. Listening is, therefore, an act of exposure that has the potential to both soothe and disrupt, involve and alienate. The shift to thinking about listening as an applied research tool requires us to think about the "conditions of connection" (Stewart 2007: 31). These are the landscapes, netscapes, rooms, and roads where we listen and are listened to and where some voices go unheard or are erased or distorted.

Listening and Story

The applied research discussed in the book stresses the importance of listening to individual narratives. This, in turn, has prompted me to ask questions over story and its performance within the narration of the self; across digital storytelling and other projects asking for life-course narratives. I don't think, like Stead or Kassabian, that there is anything problematic with using narrative (Macarthur 2016: 175), on the contrary, continuing to use this "narrative paradigm" (Macarthur 2016: 175) meant that participants "knew the score"; they understood what "story" meant to them. A lot of the book will be showing how it is the stubbornness of story and its desire to impart order, to emplot (Ricoeur 1984), to frame experience and emotion that is being listened to across a number of different research "listening spots." Story is integral to the pursuit and presentation of memory. Kuhn argues that "one central plank remains however: the notion that memory and memories are discursive" (2007: 283). They are iterative, and the past is only available through this process, whereby it is "iteratively reworked and enfolded through iterative practices" (Barad 2010: 260).

A number of these stories that are discussed in Chapter 3 were rehearsed, scripted, and repeated in ways that have meant that I use the phrase "performative historiography" to capture how the repeated narrations of the past self incrementally become an accepted version of that same self, broadcast to others. I remain convinced of the pull of narrative, "canonical linguistic frameworks that organize event memories into comprehensible and causal sequence of events in the world" (Fivush 2008: 51). This organizational process can be accompanied by a similar process of the "ensemble and sequencing of images" (Kuhn 2007: 286) for both individual and collective subjectivities has been a feature of the empirical digital story projects I have worked on, and they are too, a feature of the online archives I explore in Chapter 3. They are all connected by narrative, and these narratives lie at the micro, individual level. I have tried to spotlight narratives that get subsumed under larger stories, to give people the mike.

Not Listening

In 1980, the British cultural theorist Stuart Hall introduced an idea about encounters with media messages. His famous "encoding, decoding" (1973) article illustrates the vectors of power that lie at every point along the trajectories of (then twentieth-century) media messaging. Depending on cultural capital, status, education, age, and political affiliation, viewers might decode messages that producers had encoded into them in one of three ways: dominant, negotiated, and oppositional. This model allowed for an understanding that "reception" might be colored by a number of different variables, as well as relying on a substantive difference between the then "producer" and the "audience." The model worked by understanding the ways in which television and mass media were encountered; I watch the evening news and disagree with its coverage of an event due to my politics; my family agrees with it— and so on. Put listening in the model. That doesn't change much on an initial attempt, you can listen and agree, listen, and partly agree, listen, and disagree. But how about listening and ignoring? How about switching off? Some of what the book covers in Chapter 4 is about specific institutionalized forms of "not listening" to those who have exceeded the boundaries of ordinariness by being too noisy.

Ways of Listening: Interfaces and Interactions

Listening at the door, through telephone wires, through headphones, across oceans, to old vinyl, to zoom lag; ways of listening are formed and impacted by technologies as much as subject position (Barthes 1985b; De Nora 2000; Born 2010; Bull 2004; Nancy 2007). Emerging from the applied research work that is covered in Chapters 3, 4, and 5 is an understanding that "practices of listening are also shaped by technologies and their interfaces and affordances" (Rice: 102 in Novak and Sakakeeny 2015). Listening is not only a fulcrum between listener and listened to; it is a process that collapses (and enforces) difference, allows for proximity, and happens within broader political, chronological, and temporal flows. It also operates on a dynamic axis of ever-mutating digital communication practices that enable listening across time, and across bodies. There is a wealth of research on headphone practices and the making of the personal sonic environment (De Nora 2000; LaBelle 2010; Bull 2018; Hush 2019; Roquet 2021), where the interface between the self and surrounding physical space is moderated through soundtrack choice. My interest lies at the interface of the listening self with others enabled by, troubled by, and afforded by applied research listening practices and recording technologies.

Before I go any further, I want to think about my own "interface" with listening, since that has been the impetus for the book. I acknowledge the burgeoning impact that autoethnography is having within feminist media and popular music studies (Ettorre 2016; Grist and Jennings 2020; Cohen, Grenier, and Jennings 2022). I do not specifically work from an autoethnographical perspective, that is, I am not critically examining my own responses to and experiences of listening throughout the book, but I guess I am borrowing from its spirit, what Ettore calls the "feminist 'I.'" My body has been present in the research that is written up here. And it is not an invisible one, it is not transparent. It has entered into encounters and seen its impact; its accent, title, age, nationality, and gender have consistently marked it out. It has seen these borders negotiated in these listening exchanges. But I have chosen listening because of two life experiences. First, like many, lockdown made me think about my listening in a different way, and I wanted to put it at the top of how I worked with people in research projects. Second, I suffered a breakdown in 2019 and over the course of three years, was listened to. It came at a price, a hefty one, but it made me realize how important that space was: how the act of listening is an act of kindness. And I used that experience to think more about listening as an act of defiance.

Map of the Book

Listening ripples across time and ruptures the surface of things. Bodies lived and bodies politic can be disturbed by the act; taken back, brought close to Others. Reassuring, reminding, and disturbing, listening is the connecting mode, and Chapter 1, "Connecting Lineages," brings together established and emerging voices from across new and not so new philosophies of listening that revolve around the idea that listening is caught up with resonance and refusal. I use ideas from sound studies and memory studies, from cultural geography, sound art, and popular music opening up a conceptual space to build a theory of listening in relation to belonging and memory, touching base with key work on listening (Barthes 1985b; Erlmann 2014; Kassabian 2013; Kheshti 2014; LaBelle 2018; Nancy 2007; Szendy 2008, 2017; Voegelin 2014; James 2019, 2021), listening and Otherness (Denning 2015; Radano and Olaniyan 2016), and listening to images (Campt 2017).

Chapter 2, "Applying Connected Listening," discusses how to listen with an ear to memory and belonging. This methodology can be used across interdisciplinary research projects, including, but not limited to, digital storytelling, popular music ethnography, online listening, and media analysis. It can be a messy method, given that it involves bodies and entanglement, upset and discomfort. I introduce the term "Grounded Feminist Listening" as a negotiated way into this type of research listening. This approach relies on listening as a "feminist care" mode, particularly in terms of "giving attention" to (Robinson 2011; Sevenhuijsen 2003; Brannelly and Barnes 2022; LaBelle 2018), which in turn, can produce discomfort through unforeseen affective connectivities (Stewart 2007; Kheshti 2014; Schulte 2016), and double witnessing (Dreher 2009; Fernandes 2017; Hirsch and Spitzer 2009). This is when narrators and listeners witness unfolding stories triggered by artifacts and images in parallel. And because the research involves listening to images, I use Campt's (2017) work on the resonant image to understand what this means in practice. This includes using "Qualitative Online Listening," a simple modulation of traditional media discourse analysis, which I use to tackle the resonant image and its affinities, or "intensities" (Stewart 2007) across applied research, with participants and with online narratives of the @ AuschwitzMuseum account.

Working with veterans, migrants, rural workers, farmers, and people being helped into employment has made me consider the ethics of listening across different societal layers; how might we listen across class and across gender, and what might a listening with age "look" like? What are the ethics of me, someone who has never been in care, or in prison, or has families who were murdered in the Holocaust, or who have been trafficked or fled war, listening to voices of those who have? What are the ethics of "curating" those voices to gauge the impact of policy implementation? (Butterwick 2012; Dreher 2009; Fernandes 2017). Some of the conclusions of this chapter reveal more problems than answers and these are further explored in Chapters 3, 4, and 5.

Chapter 3, "Listening Across Age(s)," is about listening to the past as it is "storied" in the present. It introduces the idea of "performative historiography," which is a process whereby stories about the past become "true" in the repeated telling and how some of these stories and the formats they appear on can go some way toward upsetting dominant narratives. Because it is about listening across age, it also introduces the idea of "rippling," which thinks about listening as multi-temporal (Barad 2010; Tanner 2021).

The chapter focuses on two case studies. The first uses a digital storytelling project, Veterans' Voices, where I worked with aging UK veterans over a two-year period to facilitate intergenerational storytelling and listening. I use memory research with war veterans (Islam 2019; Long et al. 2021; Nugin 2021; Welzer 2010) to help make sense of the project and consider how listening across ages actually manifested. The second case study listens to the Twitter account "@AuschwitzMuseum" using work on archives and witnessing (Felman and Laub 1992; Hirsch and Spitzer 2009) and memory and image (Kuhn 2010; Tendeciarz 2006). Every day, the account, called 'Auschwitz Memorial', posts twelve photographs of people who perished in or survived the Holocaust. I argue that this listening is an enfoldment into an affective economy (Ahmed 2004), which provides affective cross-temporal connections (Stewart 2007). On October 27, 2022, Elon Musk bought the platform, which resulted in staff departures and forebodings of platform collapse. If this happens, then that listening will have to move elsewhere and its concomitant affectivities may alter.

Chapter 4, "Listening and Belonging," explores how modes of listening work to generate ideologically specific ways of belonging and "not-belonging" as a citizen. It discusses this from a UK perspective, using the early twentieth-century radio show, *Listen with Mother* (1950–82) and the recent "The Listening Project" (2012–22) which offered "ordinary people" the chance to have a conversation. I illustrate how this urge to the ordinary (Stewart 2007; Berlant 2011), a way of being, an epithet, that can be both inviting and constraining, is mobilized to shape who is listened to and who is excluded. Moving to recent case studies from recent UK legislation around protest and noise I pick out instances where there are those who are not listened to and threatened with expulsion from this good "citizenry." Through discussion over a digital storytelling project with unemployed people, it then asks questions over listening as a way of gauging the impact of routes to good citizenship that are couched within a neo-liberal framework of the good citizen as the working citizen.

Chapter 5, "Listening: Migration, Voice, and Place," is about listening in order to question certain "western econom[ies] of voice" (Chouliaraki and Georgiou 2022). It uses a two-year pan-European project called Mapping the Music of Migration (MaMuMi) where migrants in Bulgaria, Cyprus, Greece, Italy, Spain, Norway, and the UK shared a story about a song that was important to them. Its aim was to counteract dominant European-wide discourses of the "migrant" through listening exchanges centered on a song. The chapter argues that listening and its ability to produce proximity (Erlmann 2004: 175) and "affinities" (Stewart 2007) is an affectively powerful mechanism. It can make the silenced subaltern (Spivak 1988), audible and engineer a space for sonic agency (LaBelle 2018) for those whose voices have been collapsed into prevailing and pervasive discourses of Otherness, fear, threat, and vulnerability (Chouliaraki and Georgiou 2022).

Chapter 6 is called "Echoes" and is an attempt to resound some of the key ideas that emerge throughout the book. It adds in some more voices and asks you to listen to them. One of those voices is Robert Landberger's.

1

Connecting Lineages

What map are you in the process of making or rearranging? What abstract will you draw, and at what price, for yourself and for others?
                                        Deleuze and Guattari 2013: 238

Deleuze and Guattari ask a good question. The academic assumes the role of the cartographer. What a job, what a duty, what responsibility. What a deal it is to try to carve out a conceptual space that allows them to speak, just for a moment, about what they want to contribute to that map, what to build on and acknowledge or ignore. Where should they cut, where should their borders be? Mapping is about control, maybe even violence, it is about owning temporary territory, setting perimeters, and sealing off via boundaries and borders. It is about silencing others that are outside of it (of which more in Chapter 5). All maps are instruments of power (Barnd 2017; Fraser 2018) and involve an understanding of who can and cannot speak about a subject. Making a map to ground your own contribution so that you can speak with some element of authority about it means making not just a mental or physical projection (Fraser 2018) but a productive one (Mason-Deese 2015). Who should I cite? Who gets a mention and who is left languishing, and do I really have to refer to Adorno again? I don't, I choose not to. I've heard enough. Let's bring others into the discussion. The price of this is a turning toward including more recent contributions from more contemporary scholars working on listening and sound (Kheshti 2015; LaBelle 2018; James 2019; Finer 2021; Donovan 2019; James 2021) and situating my understanding of listening as something that happens in micropolitical moments of enunciation (Bassel 2017), where listening is part of the triad that includes belonging and memory; the three intricately woven in a constantly dynamic negotiation as they are experienced and encountered.

To do this I need to note that certain hegemonic assumptions have come to dominate mainstream sound studies, particularly around a perceived audiovisual binary, which I outline in order to clarify the field as it appears, but my conceptual map also involves those whose contributions do not ricochet around the walls of entrenched academic corridors of power. I make a play for newer voices. I do this precisely because my work is about listening to stories voiced by those who have not told them before and to people who have been and continued to be made mute by political discourses and dominant narratives. So rather than starting with an established canonical voice, I want to begin by using some of Ella Finer's words, written in 2021. Working in sound installation, and writing on feminism and listening, her ideas on listening, time, and witness echo mine and I want to (re)sound them to start things off:

> We travel far-away by ear, and we hear what's outside our own window in sharper audibility, the range of our hearing expands. And with this new awareness, is an acoustic attention to the interconnected world. And our place within it. Some place microphones to hear what is around us, beside us, within us—right now. The streamers. By opening sound channels across the globe, these inaudible bodies remind us that listening happens in the present tense that always holds the fragile past and future; they remind us that listening is to witness, and advocate for, the world as it changes; that listening is a practice of action. (Ella Finer 2021)

Listening is a doing thing, a "practice of action," it is something that I have been doing with and among different people on the various projects that I write about in the following chapters. Through my practice, I have come to consider concepts of listening and have stumbled across a huge terrain of contributions. I set these out here for what they allow me to think about the work I have done, and for how they situate that work within certain conceptual trajectories.

This chapter is therefore, necessarily, a selective mapping exercise, an effort to build my own *strategically essentialist* working platform from which arguments might be made around listening, belonging, and memory that can provide some insights into the relationship between all three. It is a map that has come to make sense as I drew all the projects together, wondering how to theorize them. And in so doing, I have made a temporary pontoon constructed out of selected contributions from cultural philosophy, sound studies, and memory studies to frame, or "undergird" (De Nora 2000: ix) my work on listening, belonging, and memory, and to foreground their interconnectivities. Inevitably, there will be gaps, but this review is about putting different bodies of literature into dialogue with each other, to bring them into contact to see what new associations might emerge. I also hope my selection disturbs by providing a different connective theoretical "tissue" that might be applied to thinking about listening across applied media and popular music research.

The chapter title uses the word "lineage" to suggest the multivalent and tangential lines of inheritance that each of these three areas mobilize. Each line would constitute its own review chapter, if not an entire book, each discipline has its canon. I include some key figures from those canons, notably sound studies, because they were new to me, coming from popular music studies. This is not a textbook, a primer on sound or memory studies, it is a critical but subjective review based on qualitative research about the connections between listening, belonging, and memory that comes out of and informs my research. I have gone some way to "making up" a new map, introducing new and/or unexpected alliances and by doing so, trying to open up fresh commonalities and connections. I am putting these bodies of thought into dialogue with each other because I have needed to; the work I have done with people and their stories, in person and online, has made it necessary. I listened to people, their stories, them speaking about songs, their silences and I had to consider, through recourse to the literature available, what I might be doing. I knew that I was not talking about listening in relation to sound, or music, but a listening that was pegged to story.

The small acts of listening I have been involved in are politically and individually, affectively, vital. They exist and matter in a way that Couldry, writing on digital storytelling's potential said in 2008, how it "is part of a wider democratization, a reshaping of the hierarchies of voice and agency which characterize mediated democracies" (2008: 11). They have happened in micro spaces that allowed for small bubbles of exchange whose very existence has been testimony to some "rustling and stirring" (LaBelle 2018: 61). Sitting within larger superstructures, they are "creative engagements" (Western 2020: 294), part of what Hrycak and Rewakowicz call "micro publics" (2009), or "micropolitics" (Bassel 2017: 9), affording moments of sonic agency (Denning 2015; Radano and Olanyan 2016; LaBelle 2018).

The first area that I map out as a discrete block is the act of listening. This is arguably a large and potentially unwieldy arena, so I have set its parameters to include some of the influential works from sound studies and philosophy for what they offer in terms of understanding listening's relationship to the embodied self and its place in space and time (Barthes 1985a; Clarke 2005; Nancy 2007; Szendy 2008; Kassabian 2013; Erlmann 2014; Voegelin 2010, 2014, 2019). These writers have offered definitions and distinctions (Barthes 1985a), have reconsidered its history in relation to a perceived visual dominance (Erlmann

2014), have defined resonance in different ways (Nancy 2007; Erlmann 2014; James 2019; O'Keefe and Nogueira 2022), of which some interventions mark out the relationship between visual culture and a denied auditory Otherness (James 2019; James 2021). This segues into a section on listening across and uses ideas from popular music and sound studies to think about how sound and listening sit within and across colonial borders and experiences of empire (Kheshti 2015; Denning 2015; James 2019; Radano and Olanyan 2016; LaBelle 2018). This works to ground the ideas I set out in Chapters 4 and 5 that circle around the politics of sonic control, and of muting.

The final section is about listening's relationship with the past. This involves thinking about how it operates as an act which Finer says is "in the present tense that always holds the fragile past and future" (2021), or something whose presence is tinted with past and future traces. Bergson (1911), Schwartz (2005), Barad (2010), and Hristova (2020) are included for what they have to say on this relationship, but I do spend more time with Kuhn's 2010 work on image, memory, and narrative, because I have used her approach. For Kuhn, photographs or images remain key to the narrative process, and these images become resonant and are not just seen but listened to Campt (2017). Being brought into dialogue with images and narratives is to be placed as witness, and this emerges as a key concept within memory studies of the Holocaust particularly. Canetti (1979) calls the listener an earwitness, which Birdsall, writing on sound memories of the Nazi period, refers to as "a key mode of experiencing and remembering" (Birdsall 2009a: 170). They go on, "Canetti also sought to highlight both the selectivity of remembering and the amplification of certain memories over time" (p. 170) and this mutability of the listening/witnessing process is noted by Hirsch and Spitzer, who argue that "a complex relationship between enunciation, listening and truth" lies at the heart of memory studies (2009: 152). This complexity is foregrounded in what I have come to call performative historiography and performative listening, whereby patterns of listening are repeated and performed within shifting but recognized aural and visual models or templates, and these patterns are subject to Canetti's selection and amplification.

Types of Listening

The urge to define is, as Foucault knew, about control; it's another form of mapping, shutting out, firming down, making clear what is and what is not allowed and acknowledged. The taxonomic imperative to label is in part a way of trying to distinguish a "thing" in order to be able to approach it. And so, listening has been defined in opposition to what it is not; listening is not hearing, listening is not seeing. I have had to define the kind of listening that this book is working with as I encountered the kind of listenings that were happening across the research projects with veterans, with migrants, and then with how it was being used by government. To do this, I needed to find out what was out there in the field already and to converse with these different approaches in order to inform and contextualize my own position, which is that listening to voices telling stories is always contingent on the trio of variables of subjectivity, place, and memory. Listening is an act, but it is also a connection, and operates, like the nomadic subject (Braidotti 2013; Deleuze and Guattari 2013), in a "complex" field of interrelationships, human and non-human (Macarthur 2016: 7), shifting in response to the listening matrices within which it finds itself. Listening happens on a "spectrum" (Gallagher 2017: 622). It is not "neutral, it always involves a particular type of ideological and practice intervention" (Marsilli-Vargas 2014: 42). It is learnt, taught, and expected. To listen wrongly or to not know how to listen is not just about appropriacy, that is, what is the right way to listen, in what instance, in what place, it is political. Listening is classed and raced. Listening is gendered. Listening is about regulation. Macarthur writes how "Listening, along with hearing and the 'ear', turn out to be metaphors for a complex constellation of physical, psychological, historical, and cultural factors" (2016: 171). This complexity is revealed when we start to see how many ways into thinking about listening there are.

Listening takes place in contexts, in and across bodies, time, and online, and so there are many types or "genres" of it (Marsilli-Vargas 2014). A list of listening types might include, but not be limited to the following: acousmatic, expanded, deep, emergent, close, psychological, physical, extended, immanent, reflective, and reduced. They exist as different modes, intensities, or expectations, and are contingent on location or intent (Szendy 2008), a three-way encounter between sound, ear, and place. Listening is produced in locations too and different spaces produce different listening opportunities: listening alone, online, in an office, listening in a group, assembled for a screening of collected memories, outside at a rally, inside for a concert. To illustrate this, I want to describe somewhere a specific place where I listened on two, very different occasions.

On Monday, June 8, 2020, an event was held in the Regency era park of Cheltenham Pump Rooms. The building was central to Cheltenham Spa's

Regency popularity and was the place where the Monarch of the time, George III (in 1788), went to "take the water." Colonnades, white stone, facing down to an ornamental lake across sweeping green lawns, the Pittville Pump Room has also been a classical concert venue. I went there in the early 1980s to watch a string recital, as part of a school GCSE music trip. We kept silent and applauded when appropriate (looking to our teacher to see when that was). We were learning how to listen to that type of music (baroque) in that type of space. Local Black speakers and poets, mostly young, took to the bandstand just in front of these recital rooms, speaking up in outrage at the recent murder of George Floyd, who had been killed by police in Minneapolis in full public view on May 25, 2020. We were in the first lockdown of the Covid pandemic and were carefully spaced out from one another. We were masked. American hip hop was playing over the speaker system, drifting down the grassy slopes to the ornamental lakes. Everyone took the knee for eight minutes and forty-six seconds, the time it took for Floyd to be murdered as police knelt on his neck, and I listened to that silence.

The fourteen-year-old listener was part of an audience of a baroque recital in the rarified surroundings of a classical concert in a well-heeled provincial English town. The 56-year-old listener was part of an audience of listeners who were peacefully protesting. As Stead argues, "Listeners are therefore produced through the different mechanisms and manifestations of power by way of the institution" (Stead 2016: 180). The space within which the listening is being done is already marked with expectations, rituals, and histories of in and exclusions. The material and sonic architecture of the 1980s' recital produced a different listener than the #BLM event, because listening is dynamic and a node within a network of other connections: historical, personal, and environmental. The kind of listening that was expected of the young teenage (white) me was one born out of Western classical traditions of musical listening, and I was privileged to be in the position of learning it. The listening I was doing on June 20, 2020, was a sympathetic listening with, as well as an empathetic listening across. It was a painful listening, born out of sadness and empathy and a sense of duty as a witness, and the two types of listening that I did forty years apart in the same place indicate the complexities and contingencies of those constellations

For now, I am not interested in the material or digital technologies of listening, which as Georgina Born notes have undergone "profound transformations" (2010: 85) over the past century. The applied research that I am reflecting on here is about the interplay between listening subjects and objects. I am also not dealing with "ubiquitous listening" (Born 2010: 8), the kind of everyday listening that "fills our days" (Kassabian 2013: xi) nor the "metropolitan" or "social media" listening (Radano and Olanyan 2016: 3, 4) circumscribed by new technologies. I do not engage with theories of musicological listening beyond a brief note of those that consider distinctions (Anders), or environments (Kassabian 2013; James 2019); that is, I do not engage with Shaeffer's four modes (1966) or Chion's audiovisual listening modes (1983) because my focus is pitched toward listening as a connecting process linking people and stories through voice. I think about listening in relation to the heard and unheard, about silence and not listening. In a paper on the taxonomies of listening (in relation to understanding) Tuuri and Eerola (2012) talk about nine listening modes, which they list as reflexive, kinaesthetic, connotative, causal, empathetic, functional, semantic, reduced, and critical listening, and these "refer to different constituents of meaning-creation in the process of listening" (p. 137). The research and analysis in Chapters 3, 4, and 5 actively use kinaesthetic (in listening to the materiality of images), empathetic (in carving out listening spaces and sharing stories), and critical (in Chapter 4, on citizenship and the use of sound to police the borders of belonging).

You cannot move thought to the ear without a struggle, much of which is based within a hegemonic grip given over to the primacy of the look, where the binary seesaw of vision = distance = accepted on one side and aurality = proximity = denied on the other remains a dominant force to be reckoned with. And so, the next section is a short review of some major interventions on the topic that hover around this binary, moving through those schisms and splits toward a more generative and expansive view of listening, one of interactions and interstices, of potentials and pauses, of meaning made in gestures and intimations, reflections, and repetitions. It starts off with a personal reflection on that split as it is played out in contemporary health care in the UK.

Overcoming Distinctions

Three children. Three babies. Three fetuses. Many scans. Each time the scan peered through maternal skin and flesh to view "baby." Picture taken in black and white and shown to relatives as an assurance. The scopic regime in action. We saw it so it is there. What about me who had listened to the three in the womb? Hearing movement, sensing somersaults. A pregnant listener is not enough. Maternal listening is not an official documentation. The state comes in as health-care provider to guarantee status through seeing.

Listening is not as valued as seeing. The body is subservient to the gaze, even while it emerges from it. The gaze, the view, the text, and the look, or as Kassabian puts it, the "Enlightenment obsession with vision and perspective" (2013: xvi) is remarkably entrenched and clung on to as a valid epistemology that Voegelin calls inherently "masculine" (2019: 159), a "central plank of modern Western societies" (James 2021: 5) allied to racism and capitalism. There are contemporary sound studies scholars who consider the body and the beyond-body as integral to a philosophy of listening (Eidsheim 2015; Voegelin 2014, 2019) but this pregnant body's ears were far from being heard in that hospital room. Seeing was believing. The event was a triumph of the visual over the sonic/auditory, a relationship to which I turn now, parsing key theoretical contributions on listening. By noting them, I am not agreeing with them, but acknowledging how they form the canonical corpus which I am including because it was new to me, and so maybe to others reading this book who come from digital storytelling, or memory work, or popular music, or feminist media studies. Barthes modal approach is important to include because it points toward a less coherent, more fuzzy way of thinking about listening.

Distinction #1. Barthes (1985a) argues that there is a split between hearing and listening. Listening is "psychological" whereas hearing is "physiological" (Behr 2014: 114). In French, this distinction is played out between "écouter" (to hear) and "entendre" (to listen) where "entendre" carries within it a sense of understanding or meaning. Barthes then further taxonomizes by suggesting that there are three orders of listening. The first is an alert, like an animal's response to audio stimuli; it happens in the body and is pre-rational. The jump cut in horror movies works on this model. The second involves deciphering audio stimuli in accordance with linguistic codes and memory, and so suggest that listening is done in context, where that context is both cultural and personal; it is, as Sterne says, "a definite cultural practice" (2003: 19). Some of these might surface in "'audile technique[s]' . . . a set of practices of listening that were articulated to science, reason, and instrumentality and that encouraged the coding and rationalization of what was heard" (Sterne 2003: 19 in Kheshti 2015: 8). This listening is done to produce meaning, to deliver sound into the *episteme*.

Barthes' third mode identifies listening as a process whereby we listen for what cannot be articulated in meaning, that is, through language.

> "the third listening, whose approach is entirely modern (which does not mean it supplants the other two), does not aim at—or await- certain determined,

classified signs": not what is said or emitted, but who speaks, who emits: such listening it supposed to develop in an inter-subjective space where "I am listening" also means "listen to me." (Barthes 1985a: 246)

This moves the emphasis away from content to subject positioning, from what is said, and what is heard, to the dynamics of the exchange; it deprioritizes meaning and foregrounds intersubjectivity. Veit Erlmann, an important contributor to the history of the debate on sound and vision, sees it as an attempt to "overcome the distinction between a language driven experience operating within the confines of the Saussurian signifier-signified dichotomy and an unmediated, corporeal experience" (2014: 21).

Distinction #2. A separate urge to distinguish listening from hearing emerges in more recent geographical work (Gallagher et al. 2017: 662) where listening is aligned to activity or attentiveness (Kassabian 2013: 110) and hearing to passivity. These distinctions may not be overly secure and Kassabian notes their potential circularity of (2013: 7–8) but my use of the word "listening" is premised on Gallagher's idea, that it is attentive. He writes how "Listening is active; it allows age, experience, expectation, and expertise to influence perception" (1989: 3 in Gallagher et al., 2017: 622). I want to keep this model of contingency and modality and use what Stead notes in his description of Cesaro's "multimodal listening," which is listening as a "full-bodied act" (p. 103).

> Rather than treating listening as something we *do*, I will conclude by reflecting on what happens if we think about listening as a concept. A concept, according to Deleuze and Guattari, acts as "a condition for the exercise of thought." Concepts are an act of creation and take shape only in accordance with other concepts and with problems. The concept of listening, then, can be understood as being tied to the problem of sound, and it will take different forms according to each iteration of the problem—it is thus fluid. (Stead 2016: 185)

Given that I am writing about listening to voices, both in embodied interactions, online and across time, to listening to the voices of one's own past, to one's own voice and to the imagined voices of others, this idea of listening as somehow movable is really helpful. If listening is something which activates something else in association with its environment and potential, then it has an elasticity to it which makes it a moment, not a "thing," something I explore in the following chapter in relation to listening across age. This elasticity is not apparent in the split between the visual and the aural that continues to exert its distinguishing force, as it did in the hospital room.

A Resounding Split?

What if this perceived split has obscured something, and that something is that all the while the gaze was thought to be in the ascendant, where reason was prime and prized, resonance was lurking around and had a part to play. Erlmann argues precisely this and clearly sets out what the split is and how resonance works within it:

> Resonance is of course the complete opposite of the reflective, distancing mechanism of the mirror. While reason implies the disjunction of subject and object, resonance involves their conjunction. Where reason requires separation and autonomy, resonance entails adjacency, sympathy and the collapse of the boundary between perceiver and perceived. (Erlmann 2014: 9–10)

So, this is the split as it has been staked out, but Erlmann suggests that there has been a greater degree of interplay between reason and resonance than perhaps may have been previously suggested and that the ear is something we think with.

He reflects on this "tyranny of the binary" (2014: 340), arguing for "knowing bodies and feeling minds" (2014: 340). The division between sight and sound is considered in his comments on the phenomenological work of Günter Anders, where he notes that "Seeing is an 'act' that confronts its object. Listening by contrast, is a form of *Befindlichkeit*, a state situated somewhere between 'act' and condition, between object-directness and objectless disposition" (Erlmann 2014: 327). *Befindlichkeit* translates as "situation" or "disposition" and articulates what Erlmann is arguing toward which is to think of listening (or as he puts it, resonance, or aurality) as a processual thing that is about placement, alliance, co-presence.

He writes how "the story most commonly offered about the making of modern rational selves as the progressive silencing of sensation and emotion as inherently incompatible is only half the story" (2014: 11). Giving examples from the history of Western philosophy between 1633 and 1928, he "consider[s] resonance as being inextricably woven into the warp and woof of modernity" (2014: 15). His argument rests on the contention that there does not have to be a battle between *acousteme* and *episteme* (p. 15) and that the "intermingling of subject and object . . .; [affords a] complex, sensory entanglement with the world" (p. 315). This entanglement might contribute to a sense of confusion and what he calls "boundary collapse," which is clearly terrifying because it imperils the Western subject's hermeticism and threatens dissolution. The leaking body is a monstrosity that has consequently been policed, not only those female ones Elizabeth Grosz talks about in *Volatile Bodies* (1994) that secrete, bleed, mutate, give birth, and are unruly, and so ruled. Erlmann's talk of collapse is to some extent about control, and the threat of unreasonable assault, whether it is couched in the body or through the ears. I look at this, the control of noisy and unwelcome bodies, in Chapters 4 and 5, as they appear in the sounds of protest and in the too noisy voices of migration, which are policed in order to mute and render silent in the pursuit of preserving a quiet, conservative whiteness.

Listening Across

Listening across can be about listening across time, or gender, or class, some of which I address in the next chapter. Here I am thinking about listening across as to mean listening to voices and stories from across raced and/or policed borders. I use popular music studies' work on sound, colonialism, and difference (Denning 2015; Kheshti 2015; Radano and Olaniyan 2016) to think about the politics and mechanics of sonic agency (LaBelle 2018). This agency, this way of speaking out and of making noise describes the mechanics and technologies through which marginalized and colonized subjects have sounded out. It is specifically applicable to the material in Chapter 4 where I use it to analyze the politics of undesired sound, specifically of unwanted Otherness in a specific urban square in London. Martin James (2015) alludes to it in his work on Black British reggae systems and James' (2019) reading of acoustic resonance makes a space for it as "sound" that refuses resonance, that is, it is sound that is not relatable, not co-optable, not couched within the terms of reference within which resonance resides.

Sound and its relationship to the politics of race is covered in James' 2019 book on sound, resonance, neoliberalism, and (largely) US popular music. This work uses a specific reading of resonance built on its sonic properties and a number of different philosophical approaches to read American popular music, to argue that the sonic episteme is not a seemingly refreshed way of knowing but sustains neoliberal biopolitics and white patriarchy. This sonic episteme "uses concepts of acoustic resonance" (2019: 3), which, she argues, "hasn't had a singular or stable meaning through the history of Western thought" (2019: 97). She does offer a description of resonance to suggest that it "describes either the consonant rational interaction of phase patterns or the dissonant irrational interaction of such patterns" (2019: 63). Resonance then is about recognition and misrecognition within a specific politics and the "phase relationship" (2019: 8) continues what she terms neoliberal biopolitics.

> Philosophers regularly present sound, voice and resonance as rescuing Western liberal democratic theory from the errors that speech-and vision-centrism have supposedly baked into its foundation. This chapter addresses the variations of this move that understands sound, voice, and resonance as acoustic resonance. I show that these ontologies are constituents of the sonic episteme because they take acoustic resonance as the basic unit of society and create the same relations of domination and subordination in that society that neoliberalism and biopolitics use statistics to create and maintain. (James 2019: 55)

There may be ways to think about the "rhythms of living" (2019: 182) that aren't "coded out of circulation" (2019: 182), in spaces of productive expression that do not collapse under the weight of co-option (see also James' 2021 work on Black British reggae). She uses Black scholars of sound (Havis 2014; Weheliye 2005) to analyze audiovisual performances by Rihanna (and Beyoncé) to suggest that "sounding" might open up a space that lies outside of the "domination and subordination": "sounding conceives of sound as an oblique pivot or queerly charged spark instead of acoustic resonance" (2019: 86). Where sound might operate as a pivot or spark, then it might sit outside the ripples of the resonant, which remains the terrain of the dominant. The "pivot" changes direction, switches us into a different space, is similar to Nancy's (2007) claim that listening opens us up to meaning. The "spark" is a jump in the fabric of meaning. The spark is singular, an eruption, the pivot is the fulcrum that I have come to think of as part of the listening process and its dynamic potential, not least to open up audio worlds, such that Denning (2015) describes. His work on "audiopolitics" is a historical analysis of a specific set of sonic and production practices that helped subaltern sonic agencies. He uses Attali to argue that making sounds has the potential to "project another world" (2015: 12), which he explores through a close analysis of the local music produced and recorded onto shellac in port cities like Havana, Cairo, New Orleans, and Rio de Janeiro between 1925 and 1930. He calls this a noise "uprising" as it introduced specifically subaltern sonic agencies in what Denning argues was a "musical revolution" (2015: 4). These "other worlds" emerge in the stories that twenty-nine migrants shared with myself and my research team in our Song Story workshops across Europe in 2020 and they did, what Denning argues. New "soundscape[s]" opened up the potential for decolonizing the ear (2015: 9, in Radano and Olanyan 2016: 26). Radano and Olaniyan explore the politics of noise and music in relation to empire, Otherness, and rationality in their edited collection *Audible Empire*. They argue that the borders of acceptance are policed within colonial audiopolitics, and that alternatives to "harmony" (2016: 8) which might rupture a Western sense of musicality are cast out as "noise" (p. 9). They write that "tonality brought into audible form a naturalized, iconic civility, which, in turn, rendered that which sounded different as many calamities of noise in need of discipline, muting, silence" (2016: 8). These "calamities," or "cacophonies," these house parties and carnivals attract heavy-handed policing (Henriques and Ferrara 2014), they arise from the Other, the Afro-Caribbean communities in the UK. One way of thinking about the nature of this "calamity" is to think of noise/Otherness/different sonic soundscapes as being too "near" by their being audible. They are too close. There is a proximity to them, as the listening folded into the sounds of the carnival, the house party and the protest is a listening that makes (pleasurable) demands on the body (see also Garcia 2015). Henriques and Ferrara focus on this noisesome-ness in their work on the history of the Notting Hill Carnival (2014).

> The MCs on the sound system make a point of inviting the crowd to "join the sound" or inciting them to "be part of it," "make some noise" and "make more noise"—until eventually somebody will hand you a plastic toy whistle to blow. (2014: 144)

LaBelle's work on sonic agency (2018) sets the scene for a type of listening that has radical potential in relation to dialogue and discomfort. He argues that "listening is marked by its capacity to instill sensitivity for what goes unheard. Listening, as Deborah Kapchan argues, enables us to linger within 'spaces of discomfort' for the benefit of exchange and dialogue" (2018: 24). These discomforting spaces might be those that provoke, or question, they might be that spark in an unexpected narrative or the segue in a song that is a new encounter. When the previously inaudible sound, like the Twitter stories of Holocaust victims in Chapter 3, or the song stories in Chapter 5, they penetrate a kind of film of expectation. Kheshti's work on world music, race, and gender continues this line of thought in relation to listening. She notes how sound has the potential to seep across. She writes: "Its sublimity penetrates epidermal, discursive, and geographic barriers—changing and being changed by these movements and reverberations" (2015: xx). This is a wonderfully evocative description of sound's fluidity and flow, which I use in Chapter 5 to think about the layers of listening that happened in the Erasmus+ project Mapping the Music of Migration. Her use of language (penetration) heralds its potential threat too. This is the language of sexual assault, and might there be a reason for this, that the sounds of Otherness are tagged to the language of rape to link back to colonial fears of Blackness and specifically, to Black masculinity. It also equates sound with the irrational (subliminal), the unconscious, which is again, a threat to the conscious (white) body politic, whose boundary is threatened.

Moving from body politic to body personal, Nancy (2007) puts the body at the center of his thinking about listening. It is a small book, but it has big things to say about sound and subjectivity, methexic (participatory) sound, and listening and meaning. He puts the body at the center of his argument, likening it to "a drum, with timbre" (p. 42–3), one where the "skin [is] stretched over its own sonorous cavity" (p. 43). Starting from this position, where the body is a sounding device, enables him to rethink hierarchies of the visual in relation to the sonic. As Toltz writes:

> For Nancy, the body is like an echo chamber. It responds to the forces of its interior and exterior. As an echo chamber, it resounds freely. When listening takes over our whole being, it opens a world in which sonority rather than the message becomes important. (Toltz, in Macarthur 2016: 192)

Erlmann also notes this, writing that Nancy's work "centers on percussion, vibration and resonance as attributes of an ego whose relationship to the world is more of a sonic than a constructive nature" (2014: 307). This has an obvious impact on our understanding of meaning, which for Nancy revolves around the idea of "referral" and resonance. He writes about the "shared space of meaning and sound" (p. 7), which indicates how it is more of spatial than a cognitive process, even though both are predicated on references and referrals, like a never-ending circuit:

> Meaning consists in a reference (renvoi). In fact, it is made of a totality of referrals: from a sign to a thing, from a state of things to a quality, from a subject to another subject or to itself, all simultaneously. Sound is also made of referrals: it spreads in space, where it resounds while still resounding "In me." (Nancy 2007: 7)

The spread of sound, its ability to seep into spaces, forge through barriers (and so akin to leaking) is something that Nancy elaborates on in relation to the visual, which he argues is "tendentially mimetic, and the sonorous tendentially methexic" (that is, having to do with participation, sharing, or contagion). He acknowledges that the two might intersect (Nancy 2007: 10) but there is something radical in his use of the term "methexic." This positions sound as something that can't be contained. It lends it a radical flow. Sound can spill out, spatially and temporally. It may be unruly, and certainly, this sense of the radical trespass quality of sound was one written into the proposed Police, Crime, Sentencing and Courts Bill 2022, which the UK's House of Lords voted down on January 17, 2021, in opposition to the Conservative Party's proposals.[1] It is hard to shut down sound, to keep it at bay. Perhaps the UK government was unwittingly in agreement with Nancy's argument that while "visual presence is already there, available, before I see it, [] sonorous presence *arrives*—it entails an *attack*" (p. 14), one which is unwelcome. Nancy develops this line of argument by writing: "To be listening is to be inclined toward the opening of meaning, hence to a slash, a cut in un-sensed (in-sensée) indifference at the same time as toward a reserve that is anterior and posterior to any signifying punctuation" (Nancy 2007: 27). Listening is a practice that opens up the world, and the idea that it might "be inclined toward the opening of meaning" underscores its potential to generate radical and oppositional spaces of encounter.

Nancy's work might be used to think about how to listen to others, and to "be inclined toward the opening" not only of meaning but of difference. It lends itself to the kind of work that Les Back's sociological *The Art of Listening* does, whereby it allows a "compassionate, sympathetic and /or empathetic mode of engagement" (Back 2007 in Rice 2015: 100). And indeed, it is this idea of engagement that offers a way of thinking about the kind of listening that happens in this book; being engaged implies a degree of being with, a spatial and temporal sharing. When we listen, we might be "listening to the beyond-meaning (l'écoute de l'outre-sens)" (2007: 31), what Nancy describes as our "turn [ing]" away from the signifying perspective as a final perspective" (Nancy 2007: 31). For Nancy, listening offers a different way of understanding the world that is not linked to the finality of language. For him, "My signature is not myself" suggests that there are other audible ways of being witnessed and of being in the world. Sometimes this state of being which listening produces is one that is about being among and between things, be they people, space, or time.

---

[1] https://www.bbc.co.uk/news/uk-politics-60032465

In-between Listening

In her 2019 work on the "political possibility of sound," Voegelin builds on some of the works of Braidotti, Irigaray, and Barad, to claim that "listening [is] a creative engagement in the between-of-things" (2019: 12). This allows her to move beyond an anthropomorphic conception of the listening self to think instead of the "co-relational between-of-things and from the between-of-subjects-as-things" (2019: 47). This shift toward the interplay between and the dynamism or latency within those in-between spaces has great potential to be thought through in relation to practices of listening. In an earlier work, Voegelin writes:

> Listening [i]s an innovative and generative practice, as a strategy of engagement that we employ deliberately to explore a different landscape other than the one framed by vision, and it is cultural vision that I refer to here, grants us access to another view on the world and on the subjects living in that world. It shows us the possibilities of sound, that which could be, or that which is if only we listened. (Voegelin 2014: 12, see also 2010, 2019: 47)

In the same work she argues that "listening affords us a different sense of the world and of ourselves as living in this world; it affords a different relationship to time and space, objects and subjects, and the way we live among them" (Voegelin 2014: 10). The "amongst-ness" of listening is what I want to keep from her work when thinking about methodologies and to listening across time (Chapter 3) and is something that Bull also explores as a kind of "being with" (Bull in Erlmann 2004: 188). The idea of being "among" leave the distance/proximity model behind and embraces circularity, a muddiness, messiness, enfolding co-location, however temporary that may be, what Kassabian calls a "distribution." For her, listening is a "range of engagements" (2013: xxi) which "engage[s] us in sensual and sensory affective processes to situate us in fields of distributed subjectivity" (2013: xxiii). She notes listening's fluidity, its connectivity:

> Listening generates place, the field of listening, continually from my hearing of myself within the dynamic relationship of all that sounds: the temporary connections to other listeners, things and places, as the contingent life-world of my listening intersubjectivity that hears the actual, the possible, and even the impossible participating in the ephemerality of the unseen. (Kassabian 2013: 3)

She rejects narrative and narratology to explain listening, and proposes instead a model for the affective "how" of listening, where that "how" happens across and between bodies (p. xxix). Her definition of affect is that it "is the circuit of bodily responses to stimuli that take place before conscious apprehension" (2013: xiii), and it turns to the body and impressions. Thompson writes how it is "often understood to be synonymous with force, or forces of encounter" (2017: 45). Later in the book I use Stewart's (2007) work from *Ordinary Affects* to further clarify those affective intensities and moments of exchange or collision that listening affords because that approach is geared toward thinking about the dynamic and unexpected relationship that listening can produce.

Other disciplines outside of music and sound studies have theorized listening in relation to embodiment and affect. Gallagher, Kanngieser, and Prior (2017) are human geographers and make the case for "expanded listening." They use the term in relation to "sound," which they distinguish as either "an object, wave, or event" (2017: 620). They argue that this form of listening "refers to the ways in which bodies of all kinds—human and more-than-human—respond to sound" (2017: 618) in responsive and sometimes affective and embodied ways that are "untethered from cochlear reception" (Scrimshaw 2013 in Gallagher, Kanngieser, and Prior 2017: 619). This has the potential to "reveal [. . .] things that are not available to other senses" (2017: 620).

Their contention that "expanded listening attunes to sound's capacity both to connect disparate bodies (LaBelle 2006, 2010) and to change them (Kanngieser 2015)" (Gallagher, Kanngieser, and Prior 2017, 620) is something that I came to realize, over the course of my research, was happening. This way of listening is kinaesthetic, it is about feeling, it is empathetic, it is about caring. This is being played with by contemporary sound artists such as Kate Donovan, who work with this idea of expansion:

> the notion of listening is used in an expanded sense, and as such is posited as a way to pay attention to—and foster care and respect for—not just the voices, sounds and gestures which often go unnoticed, but also the possible histories. (Donovan 2019)

Donovan is working in experimental radio and listening, thinking beyond the Anthropocene, and although my work is with human voices, her emphasis on expansion and the spaces that might be brought into consideration via such listening helps me look to at listening as a methodology of care, which I consider in the next chapter.

Generative Listening

Expanded or expansive listening acts as a juncture or turning point within a network of exchanges. It is part of a larger communicative "knot" (Ingold 2015), not necessarily linear. It is "something we do as an embodied process and it is not something that is able to be detached from our other sensory experiences" (Stead 2016: 184 in MacArthur et al., 2016). It flows through and across us, moving sound into and around the body and that body both into and out of its present space. Just as listening to the past, via stories or images, enables the body to switch temporal registers, so listening across, being open to new or unexpected sonic encounters enables the body to move beyond its own barriers and connect with a matrix of relational sonic aspects. Eidsheim's (2015) work on singing and listening concentrates on this type of relational listening (2015: 181) that involves the whole body; they term it "ethical listening" (2015: 150). Like the work coming out of human geography, they think of the spaces in between the listener and that which is being listened to, writing that "singing and listening are better understood as intermaterial vibrational practices" (Eidsheim 2015: 3) that can "move and define [. . .] the practitioners" (2015: 100).

These contributions on listening, what it has been, can be, is, and is not interrogate the multifarious levels and scenarios upon and within which listening happens. Out of all of them, I like Salome Voegelin's idea that "The action of listening produces the thing listened to and the listener all in one move. . . . There is no gap, nothing is hidden, the encounter is all there is" (2014: 110). The notion of the encounter, which I expand on in the following chapter on methods, allows for listening to be thought of as processual and contingent. This does not bar a listener from operating a strategic essentialism in the listening process, but it renders listening as a productive and connective node within an ever-shifting matrix of listening points. These points are across and beyond the body. Listening is trans-corporal, trans-epidermal and it can be frightening. If it is a "a continual openness to the not-yet-known [. . .] what Davies describes as 'emergent listening'" (Macarthur et al. 2016: 174), which might fold the listener into others' stories and narrated histories.

Listening to the Past

I return to the voice message on a landline phone in a Dorset cottage in 2004. This played the voice of my deceased father. It was the passed in the present.

It was the past in the present. It was also now, which is now then—I turn now to consider how those interventions on listening as a process, entanglements, connection, and dynamics might work alongside the activities of listening to the past. There are questions on how we listen to the past as it is performed and brought into being in the present. As Schwarz argues, and I agree, "In memory, past and present are compressed" (2005: 141–2) and this compression brings the listener into a proximity as they listen to stories about the past told now, because listening can happen across "temporal distance" (Freeman 2010: 272 in Radstone and Schwarz). If we rephrase Sloterdijck's question "Where are we when we listen to music?" and ask "Where and When are we when we listen to the past (in the present)," then we too are compressed, enfolded, brought in. Bergson pitches the past and the present against each other in terms of how they live for him: "My present is that which interests me, which lives for me, and in a word, that which summons me to action; whereas my past is essentially powerless" (2011: 176). This is then separated from the present by what he calls "degrees" (p. 176). I would say that this is not what I experienced either in Dorset in 2004 or with my earwitnessing of the Veterans' Voices stories and the @AuschwitzMuseum account where the past both collided and collides with and informed and informs the present. These pasts are narrated and listened to, invoked, revoked, re-articulated, and relived in the present.

In an introduction to a special edition of "Memory Worlds: Reframing Time and the Past" Hristova, Ferrandiz, and Vollmeyer (2020) consider the relationship between time and memory, using Lefebvre and Karan Barad. Lefebvre's (2004) "theory of moments" work in *Rhythmanalysis* suggested that time needed to be approached "in terms of change and cycles, or rhythm". Like space, "time only attains meaning when being 'lived'" (Hristova, Ferrandiz and Vollmeyer 2020: 780). This must mean that time is socially produced and articulated. It is contingent. Encountered by subjects in different contexts, living in different time regimes, it exists within a relationship with those whose lives it might be said to circumscribe. We can go back to Deborah Moglen's idea of trans-aging, where the past is experienced in the present in a "moment" (Lefebvre 2004) to continue looking at other work on time as non-linear and think about how this might impact on the nature of listening in practice. Barad and Rovelli, whom Hristova, Ferraándiz, and Vollmeyer also refer to, think about time as non-linear. Barad's key point is that the past, present, and future are "enfolded" (they use the term "spacetimemattering" to convey this), so "there is no past moment that can be *represented* in the present, the past can only be *performed* in the present" (Hristova, Ferrandiz and Vollmeyer 2020: 779). This happened with the veterans in Gloucestershire; we were all involved in listening to the past as it is recalled, reiterated, scripted, and shared with veterans. Lefebvre's rhythms become apparent here as there is, as I will detail, a linear repetition to the telling and listening to their stories as a repetitive declamation of the self. This repetition of a rehearsed script of the self is part of a memory act that is not so much a reflection on a past time but "an enlivening and reconfiguring" of the past self that is told through linear narrative (Barad 2007: ix in Hristova, Ferrandiz and Vollmeyer 2020: 779), even though memory confuses linearity, as it "disrupts any uni-directional movement from past to present to future" (Schwarz 2005: 141).

Listening, Narrative, and Memory

Kuhn writes that memory is a "process, an activity, a construct" (2010: 298) that has story, or "narrative" at its center:

> It is impossible to overstate the significance of *narrative* in cultural memory- in the sense not just of the (continuously negotiated) contents of shared/collective memory-stories, but also of the activity of counting and telling memory-stories, in both private and public contexts—in other words of *performances of memory*. (Kuhn 2010: 298)

These stories are conveyed through "performances," and so presume a listener. And, just as the "score constitutes the listening experience" (Stead 2016: 180), so the listening experience is constituted by the memory narrative, however, it is articulated or performed. Kuhn argues that "memory has social and cultural, as well as personal, resonance" (p. 298) and once again here is that word "resonance," and now it foregrounds the idea that memory work has an inherent audibility, and so it can seep out, echo across, that it might be *heard* by a listener(s). Stewart's poetic (2007) work on affective objects and experience also considers this, writing how "objects settle into scenes of life and stand as traces of past still resonant in things" (Stewart 2007: 56). If memories and the material objects within which they travel have resonance, then their sonic frequencies will hit real bodies and have real impact, and this will be felt in those lone or assembled bodies, which became evident in the Veterans' Voices and Song Story workshops detailed in Chapters 3 and 5. What also became apparent was how "Memory . . . is not the ally of history, but its contrary. Its fallibility is the reason why historiography is required" (2005: 140). Historiography is the writing of history as narrative, largely by those who did not live out those histories, historians. In the Veterans' Voices project, however, I was asking the participants to act as historians of their own lives, something which asked them to use their memories, and others' memories of them, to script their stories. There were two things I noticed about this telling and re-telling that went on, and that was how repetition was key to the production of the stories and that this repetition was expected and performed in circular and interactive ways. Because of this I use the term "performative historiography."

"Performative historiography" is a process of narrating the self to a preset template. It involves the repetition of stories as they become the recognized written or declaimed narrative of the self. I started to think about how this might be a way of trying to capture the experience I was having when talking to veterans about their experiences in the forces, which for some participants was over fifty years ago. The way that they wrote their histories (on paper, as script) and the way they recounted them became fixed through repetition. I use the word "performative" in both Judith Butler's (1990) understanding of it as coming from J. J. Austin's *How to Do Things with Words* (1976). For Butler, it is gender that appears naturalized after repetition through resignifying practices (Gardner 2015: 30), here, in my work, listening to people tell stories about their pasts, performativity is where the saying does the doing, that is, the self that is narrated through repeated narratives becomes the past self. I also see it, as I saw it over the course of the two-year project with the veterans, as a rehearsed performance of the past self. In some cases, this self is reliant on verification from the other veterans, who might inform or inflect in turn, establishing facts, dates, places, and names, with a group consensus on what might be left in and left out of these historiographies, which were repeated in front of others, recorded, and archived in the Veterans' Voices website.

Listening to Images

In a direct riposte to the audiovisual distinction, which expects a specific route from sense to verb to object (vision/to look/painting; hearing/to listen/song), I want to work with the idea that images can be listened to. Photographs, carefully brought along to research workshops, digitized on Twitter feeds, sound out; they make a kind of noise. In *Regarding the Pain of Others* (2003), Susan Sontag writes that "Narratives can make us understand. Photographs do something else: they haunt us" (Tandeciarz 2006: 135). Again, this too-easy split between the rational (text) and the affective (image/non-text) obscures the potential they have when fused together. They are sonorous, to the extent that they have Sontag's (and Nancy's, see Toltz 2016: 192) potential to mean through sonority, but not rather than signification, more in parallel with it.

In *Listening to Images* (2017), Tina Campt argues that encounters with photographs are affective and images can be quiet or loud; they have resonance. She is able to "listen" to images through an application of Fred Moten's (2003) work that asks what "the sound that precedes the image" is (Campt 2017: 7) and in particular, by using the concept of "frequency," or "felt sound" (Campt 2017: 7), she is able to then "explore [. . .] the lower frequencies of transfiguration enacted at the level of the quotidian . . . with objects that are both mundane and special: photographs" (p. 7). It has something in common with Kassabian's "haptic image" which encourages a "bodily relationship between the viewer and the image."

What is really important about her work is that it is about listening to voices who have gone unheard, specifically Black voices in photographic archives in the UK, the United States, and South Africa. She expressly states that "My listening practices focused on the affective registers of black family photography; on how and why such photos touch and move people both physically and affectively" (2017: 23). Hers is a salvage project, a re-reading, a rescue; she attends to "the quiet but resonant frequencies of images that have been historically dismissed and disregarded" (2017: 11) and I seek to mobilize her "image-listening practices" (2017: 22) to look at and listen to the photographs that appear daily on the @AuschwitzMuseum Twitter feed, arguing that they offer alternative, personalized, narrated, and therefore affective re-articulations of arguably the most traumatic event of the twentieth century, the Holocaust. Her claim that images have resonance, that they are audible, is also hugely helpful to understanding what affective roles they play in digital storytelling, where the stories relied on listening to retrieved, rehearsed, and reiterated memories, and the memories were prompted by image.

Conclusion

A map is not just a tool for traveling. It is a resonant image, and representative mechanism, declaring importance and denial, exclusion, and power. Its perimeters veer off into silence. The mapped resonates, and the unmapped are refused. The conceptual map I have built here using ideas from listening and memory studies helps me make sense of the research I have done. In its making, I too have muted some and spotlighted others, hoping to bring a variety of different voices into dialogue.

It is apparent that through all these threads, listening has been theorized not as a thing but as a process, an affective enablement, which has run either counter to or in parallel with seeing, is distinguished from hearing, and is both in and beyond the body. Related to distance and also to its collapse, listening is the route through which moments of agency and processes of power operate and it is caught up with, and configures, belonging and memory.

I want to finish by referring back to the split between listening and use Voegelin's ideas on this to take into the next chapter. She writes: "Listening is thus not per se better than seeing; it does not prevent us from discrimination and differentiation, but it cannot avoid the responsibility in how it hears the other and the self" (2014: 110). We cannot undo the listening moment and the witnessing, discomfort, or rupture that it brings. This close collision of the self with sound, noise, story, however it emerges, is fulcrumic, contingent, contextualized, and inscribed with a variety of associated embodied materialities and hidden mechanics of power and control, of relationships with resonance. Listening is the dynamic stitch within the knot that pulls us through time and meaning, and the next chapter considers how it moves between those matrices of power and responsibility to ask how we might listen with care.

2

Applying Connected Listening

Living respectfully with other people also comes with listening. The mainstream of American society has, from its inception, been built on silencing and strategic exclusion, on not hearing who and what mounts up to the majority of us. At its most intimate it's this learned non-empathy that says that what happens to you doesn't affect me, that we are not connected, that you don't matter. At its most systematic it does this categorically: dictating that these people are not to be believed, not to be admitted as equals or participants; that they are to be laughed at or mocked or roughed up or erased.

Solnit 2020

Rebecca Solnit writes about listening in America, but her observations have relevance for researchers doing listening work in other locations too. Her argument is that there has been an endemic and systemic erasure of voices "not to be admitted," that there are some voices that are not "let in." These are the voices of the little people, the ironed out, the left behind. In my work, they are the people who live within categories: migrants, veterans, unemployed, in "care." They are always "done to," the user at the end of a government policy, the group for which some sort of solution is needed. Collected under these categories, they are rarely listened to just for the listening and its affective potential. In this chapter, I discuss how a methodology of "connected listening" can air such voices and why this is important. "Connected listening" as a methodological approach means that the researcher listens with an understanding of how this process is linked to their own body, the bodies they are listening to and the places and times they are in. Their methods of listening should be connected to their understanding of belongings and memories that the listening act affords. It also means taking time, speaking less, making space, moving over, letting listening breathe. It is a subtle shift, a slight movement of the body perhaps, and an awareness of the gaps, the stutters, the stammers, and the silences that may come.

Following Hannah Arendt's work on human existence, listening is about giving attention, and "attention is necessary not for reasons of glory or immortality, but to make real the human capacity to make our presence felt in the world" (Bickford 1996: 55). Paying attention is also about "a kind of attention immersed in the forms of the ordinary but noticing things too" (Stewart 2007: 27). The "ordinary" voices in the next three chapters are varied in provenance but are united in not having been heard beyond the domestic or personal sphere. They are the "not-yet-apparent" (LaBelle 2018: 9). They are the voices of people far from employment, often with complex and challenging histories involving mental health, crime, and substance abuse, veterans of the Korean War and of Britain's war in Aden (1963–7), and people who have migrated into and across Europe. They appear as images on the Twitter feed of the Auschwitz Museum and in conversations on the BBC archives held in the British Library. They are united by their ordinariness and in that their individuality. They are also arguably on the peripheries of power, while some of them are also at the center of media narratives around inclusion (migrants, protestors). Their stories form the material of the next three chapters in relation to age, time, belonging, and place. Here I discuss the research approach I took to listening to voices who have not been through the process of being listened to and recorded before for public consumption and use (which is different to "being heard"). At the core of the chapter is the question: What does it mean, to listen when listening is connected to belonging(s) and memory and how might we do it? What approaches might we take? What are the ethical implications, especially when "curating" those voices (Butterwick 2012; Dreher 2009; Fernandes 2017)? Where are these voices being listened to and crucially, where will they be listened to, who by and what for, and crucially, why is this listening taking place, what for? Why now?

This chapter is about how I did the listening, and how I think it afforded a different way into the research. Putting listening first opens up space, makes room, allows for a sonic agency, affords affective connections. It is just a modal shift, a slight difference in stance, but it works. It is an act of exposure, an offer of connection; it is connected listening. My focus is on four digital storytelling projects and one online Twitter account. They are about listening for empathy and commonality across markers of exceptionality or exclusion. These have involved co-temporal and co-presence encounters, in municipal offices, cafes, and function rooms, and online, because of the Covid-19 pandemic that hit since March 23, 2020. I have done another type of listening too, more solitary, not funded, not worked out with a research team. I listened to images posted daily on the Auschwitz Museum Twitter feed. Because of this, and the importance of image to the digital storytelling projects and adapted storytelling I did, I use Campt's (2017) theories of the resonant image to work out how I, and others, listened to photographs. These encounters are all linked by a research methodology that prioritizes listening as a driving mechanism for empathy and witness, one where listening challenges entrenched inaudibilities (Bassel 2022: 41). I align with Ratcliffe's model of rhetorical listening where she actions a "strategic idealism" when "listening with the intent to understand [that is] Strategic idealism implies a conscious identification among people that is based on a desire for intersubjective receptivity, not mastery, and on a simultaneous recognition of similarities and differences" (Ratcliffe 1999: 205). Brandon LaBelle calls this "sonic agency," which is a "framework for inspiring, nurturing, and empowering political subjectivity—to craft from an auditory imagination, and the experiences and promises generated from listening and being heard, emergent forms of resistance as well as compassion and care" (2018: 60). It operates as a "support structure" (LaBelle 2018: 21). Being heard, being listened to works in the same way as being seen, in so far as it "operates as an extensive psychological and affective base by which we feel ourselves as part of the world . . . *I feel myself being seen*" (LaBelle 2018: 29). It really is about taking time to involve other voices and experiences, folding them in.

Imagine what it is like not to be listened to, to be ignored, excluded, assumptions made, categorized. Connected listening is an attempt to fold new voices into a conversation, an inclusive listening citizenship. It has the potential to "shift some of the focus and responsibility for change from marginalized voices and on to the conventions, institutions and privileges which shape who and what can be heard in the media" (Dreher 2009: 445). Listening can be positively and progressively upsetting. We also need, as Dreher herself notes, to be aware of the binary discourse of marginal/privileged, silenced/silencer, and I want about lines of temporal belongings rather than essentialized categorizations. This is something that went into the design of the research. As ever, when I consider how I conducted my research, and why I chose it, I am reminded of the words of my lecturers at the BCCCS way back in the late 1990s, who asked us to think about what is at stake in our work. They were demanding that research means something, that it makes some kind of difference, makes some inroads into mapping the unheard, the under-represented: Solnit's "erased" and LaBelle's "not-yet-apparent" (2018: 9). I also needed to listen ethically and be aware of my own prejudices, privileges, and expectations (Ratcliffe 2005: 29). In her paper on the "Listening Cure," where she critiques corporate listening while arguing for a dialogic listening, Lloyd talks about the listener in the listening process. Using a more visually poetic metaphor than Hall's encoding/decoding model, she refers to Paul Carter's 1992 work on voice, where he writes, "the listener is not a pure receiver but a rugged coastline, a profile of sound history of its own, where some sounds and meanings may find a cove, beach their vessel and scramble ashore, but where others will be repulsed, and yet others misunderstood—to the point of drowning" (1992: 45 in Lloyd 2009: 484). This is akin to Davies ideas on listening whereby "when we listen, we often do so as a means to 'fit what we hear into what we already know'" (2014: 20 in Schulte 2016: 142).

The last chapter started with a quote from Deleuze and Guattari about mapping, or the process of drawing lines around, across, and between "things" in order to control. Those were lines of in- and ex-clusion of what has been and has not been deemed "knowledge" and those lines are in constant flux; I redrew them myself after learning of new work in the field and I questioned whether I had to acknowledge those whose presence has come to be unchallenged. I continue with that idea but now think about the lines that are drawn between people, notably those who exist between a researcher and research participant(s).

Some of these lines were short-lived, some lasted over two years, and they are where the listening took place and where new knowledge was produced. They connected very different bodies for very different reasons. For that very reason, I need to take positionality into account; mine is not a neutral ear nor a neutral body. It is classed, raced, and gendered; it is marked by accent, appearance, and profession. These attributes will be perceived across the listening lines in different ways. And there is a problem with positionality, an accent may mask a history and positions change, both over time and in accordance with place. Maybe we should think of our positionalities because, as Diane Fuss argued a long-time back (1989), each of us hosts a "hierarchy" of identities that fluctuate, or as Lykke phrases it, gender, and sex specifically, are "travelling categories" (2010: 40). I examine these challenges in turn and refer to work that I find useful as framing devices. So here, I reflect on the conventional digital storytelling methods I have used and adapted, consider what it is to listen to images, co-listen, and to be a listening witness. My methodology has utilized "Grounded Feminist Listening," which is based on "Care" and is about the shifting encounters between the researcher and the researched. I also refer to "Qualitative Online Listening" to justify my daily listening to the @AuschwitzMuseum Twitter feed. This is my attempt to adapt methodologies to different research terrains; it is about listening to the most sensitive and appropriate methodology for the project and the people you are working with, giving attention, and being close. And, as I noted in the previous chapter, coming close can be discomforting.

Listening and Discomfort

The four projects are digital storytelling projects with Gloucestershire veterans (Veterans' Voices, 2018–20), people in Gloucestershire being helped into work or training (GEM, 2019–22), dairy and arable farmers in Cornwall and Cumbria (the British Academy funded NetZero Project, 2022) and migrants from Africa, South Asia, the Middle East, and Europe (Mapping the Music of Migration 2018–20). They are connected by approach rather than theme. Some of them caused me personal and political discomfort, where my own politics were clearly not aligned with those of the participants, my involvement was shaped by the many factors that go into determining a funded research project. Research in a post-1992 institution in the UK exists within a matrix shaped by individual and institutional research strategies, funding opportunities, and opportunism. Projects might be hard won; they might fall into your lap. Ideally, I would want only to be working on research areas firmly aligned with my research interests, which are around women, aging, popular music, and storytelling but that has not been possible. But in those new spaces that I have been invited into, I have been afforded the space to think about new ways of listening and listening as a new way into researching.

From September 2018 to July 2020, I led a digital storytelling project called "Veterans' Voices" with Age UK Gloucestershire. This was because I was the lead for research at the then School of Media, University of Gloucestershire, Age UK Gloucestershire approached me and a colleague from the School of Health and Social Care with the view to implementing a two-year project that would open up intergenerational conversations between veterans and students. The project was part of a three-year program called "Joining Forces," an Armed Forces Covenant Fund Trust in partnership with SSAFA, the Armed Forces charity. It was delivered by twelve local projects across England by local Age UKs and SSAFA branches. My decision to take on the project came from my interest in digital storytelling and my desire to augment the research projects in my school. I had to keep my politics out of it, and when, over the two years, there were moans from the veterans about the Labour Party, for example, I did not comment. This was uncomfortable but it was instructive; I was there to listen. Here I started to think about how I was "Listening with Age." This way of doing the listening emerges from aging studies (Gardner and Jennings 2020; Cohen, Grenier, and Jennings 2022) and centers age within the listening process. What this means is that the listener embarks on the listening process within a context that is *already marked* by age, where the relationship between the listener and listened to is modulated through contextual and cultural understandings of aging. In this space listening might be performed with age; maybe deferential, dismissive. I was inadvertently being polite as myself and my participants were marked by age.

From October 2019 to November 2021, I led a pan-European Erasmus+ funded project to develop a methodology for shared listening to music. This was an adaptation of digital storytelling that had emerged from my reflections on Veterans' Voices. The Mapping the Music of Migration project (MaMuMi) was funded by Erasmus+, specifically by their Adult Education wing. I had won two previous grants from them, one to run a digital storytelling project across the UK, Austria, Italy, and Hungary with teenagers, and the other to lead a media literacy project across the UK, Italy, Greece, Malta, and Ireland. Both had been funded for around €250,000. The University Funding Office had enabled the application, as they had a long track record of successful bids and an extensive partner network. Both required that the project demonstrate alignment with EU values and aims, and the final report had to reflect on whether they had been achieved. Discomfort here came from unlikely jumps in responses, where the research encounter became more like a therapeutic one and is detailed in Chapter 5.

The third "listening project" I worked on was with a team of social economists and business development managers on GEM (Going the Extra Mile[1]), where digital storytelling was used to gauge impact on participants' stories about their involvement in the GEM initiative. GEM is a joint program involving community-based organizations and the Gloucestershire County Council supported Gateway Trust, aiming to operate a model of mentoring and support through the deployment of "navigator developers" who work with people in need of help to enable them to access education, training, or employment. Here my discomfort was at the aim of the initiative, which I outline in Chapter 4 with regard to the "good" citizen.

---

[1] https://www.glosgem.org/

The last one ran from January to June 2022 and was a British Academy-funded project around climate change and NetZero led by colleagues from the CCRI (Countryside and Community Research Institute[2]). Again, my job was to ensure that the voices of farmers and rural stakeholders were listened to as part of the discussions that were planned for two climathons, events where participants work to generate solutions to a problem, and this one was how to mitigate climate change in rural areas. Listening to farmers in Cumbria and Cornwall had never been part of a research game plan, and again, my involvement came by way of invitation. Late in 2021, colleagues at the Countryside and Community Research Centre wanted an interdisciplinary team to work on a NetZero British Academy bid and my experience in storytelling brought along a "humanities" element. The project ran for six months (January 2022 through to June 2022), with a climathon event in both locations early May. My role was to speak to farmers, councilors, and people involved in carbon neutral and traditional and regenerative farming before the climathon to collect stories about their views on their own rural heritage and their feelings about the future. I was going to be listening to people that I had never met; I was very much an outsider; not rural, not an agribusiness academic, no "skin in the game." My discomfort here was around my ignorance of the lives the people I was going to listen to led. What kind of listener would I be in these instances? What might I be trying to connect to? After a number of initial meetings where the research team (two agribusiness academics, a PhD student, and a placement student) met key actors in each area, it became more apparent that there was a plethora of voices on this subject, a number of which were in opposition to each other. The climathon was an effort to move away from debate to solutions and the stories were meant to be shown as part of them a tapestry of voices involved in those solutions. Having listened in on all the Zoom meetings setting the project up, I landed on the idea of "change." This would enable people to tell me about their lives and the challenges they faced. Five stories were recorded, three of them without me as I had Covid at the time. One of them was from a male farmer who was working in Cumbria. About halfway through his recording, he said that "I got ill and my wife left me." This was glossed over in the narrative, but it showed how listening engenders a quasi-therapeutic moment, one that is enfolded into a broader story about resilience and success, and which flavored the exchange with an intimacy perhaps unforeseen. Again, this is what a connected listening

[2] https://www.ccri.ac.uk/

approach can precipitate, because the space that you open up in the listening may only be comparable to participants "as" a therapeutic space; a chance to really say something, to let go, be recorded, and not see the listener again. This BA project recorded five digital stories and they are not part of the following chapters on age(s), belonging or migration. What they did do was offer a route into affiliation for the participants of the climathons, who, in the conversations that followed in May 2020, seemed to identify with some of the struggles that were being broadcast on farming, sustainability, and the personal costs that may have involved.

The other, non-funded research that forms part of this book is a desk-based one that I call Qualitative Online Listening. It is a moderation of conventional media and cultural studies' methods of discourse analysis, for use on the image-supported Tweet. Methodological work on Twitter published to date from disciplines beyond cultural studies and popular music studies (e.g. business behavioral science) revolves around its use as a data source (Ahmed 2019), or how people listen through it (Lee and Kao 2021). You might think that working with tweets might be ably addressed by discourse (or contents) analysis, however, I am mindful that the process of seeing these tweets over a period of more than two years and the stories they contained, necessitated a slight shift to accommodate listening, because of the nature of the stories within the Twitter space. Discomfort came here in the form of upsetting narratives, through what Stewart calls "Little experiences of shock, recognition, confusion and déjà vu" (2007: 63) that started to affect me. Appearing in my Twitter diet, which is a stream of current affairs, comments, academic news, and musicological issues, the individual stories of people murdered upset the feel of this feed. I was brought close to the micro stories of ordinary lives lost and these began to cut through the macro narratives of Holocaust history I had been taught at school. I was trying to work out what it was to "like" a post of a small girl or boy, a young woman whose early death is recorded in phrases such as "she did not survive" and have found Sara Ahmed's (2004) and Kathleen Stewart's (2007) work both helpful in clarifying the trend toward an affective economy that I, Twitter, and the stories of people murdered are part. The Qualitative Online Listening is wise to the affinities coursing through ordinary points of connection like this; it is aware of its associated affective potential and risks, and I look at it in detail in the following chapter.

Across all of these projects I have been aware of this affective sphere into which myself, my participants, and invited listeners have all been drawn by the act of listening. There is an intimacy that is often surprising, sometimes shocking. This is something to do with what Stewart calls "affective contagion" where "forms of attunement and attachment" are produced (2007: 16). I use Stewart's work on affect, mindful that the term is highly untransparent and comes with its own lineages, but Stewart's emphasis on affect and its relationship with the ordinary is helpful to my understanding of the applied research I did. She talks about affects as "not so much units of signification, or units of knowledge" (2007: 40) (in line with people like Massumi 1995), but "charges . . . [or] relays [that] jump" (p. 39) and are often "involuntary" (p. 40). This "jump of things becoming sensate is what meaning has become" (p. 42), and in these "jumps" might come whole shifts of sensibility or understanding, which is apparent in the reactions by the veterans to some of their peers' stories across the Veterans' Voices and harks back to the messiness that listening discussed in the previous chapter, where borders and encounters looked large in the literature on listening. This is because something new emerges through the encounter, and in his work on listening and children's drawing Schulte explains why. "Emergent listening 'involves, to some extent, working against oneself, and against those habitual practices through which one establishes' 'this is who I am'" (Schulte 2016: 141 citing Davies 2014: 21). Using a Deleuzian framework, Schulte's work is about how children listen, and he uses Deleuze and Guattari's 1994 work on vibration, bridges, and concepts to think about the potential for listening to "give us the courage to 'remain attentive to the unknown that is knocking at the door'" (Schulte 2016: 147 citing Rajchman 2000: 7). I like that idea of opening the door to something, someone, somewhere new, to the unexpected. It illustrates how exciting listening can be, how much of a shift it might engender. Pierre Bourdieu has written about this element of listening, arguing that the intimacy that listening provokes is both "cognitive and emotional" (in Back 2007: 94). Illouz continues this line of argument, writing of how listening is both cognitive and emotional because of its mirroring (2007: 20), a point that is core to *The Listening Project* (Chapter 4). That is, listening is about recognition; it is dialogic, a point that LaBelle also makes (2018: 18) and one that hints at the role of storying within and across bodies and subjectivities.

Digital Storytelling and Connected Listening

"Connected Listening" is a core component of digital storytelling. A digital story involves a short piece of audio edited to support a selection of still images and photographs. It is in between a podcast and a video. It is not a film, nor does it use "talking heads" or a journalistic interview technique, where questions are directed at an individual. Instead, the individual and the facilitator look together at an image or artifact, about which the participant talks. It is driven, above all else, by the voice, inspired by this response to an image, sound, or artifact. In its ideal form, it is a model of non-hierarchical co-production and access and emerged in 1984 from Joe Lambert's Story Center, where the researcher is "more likely to be that of the 'collaborator', 'curator', 'carer' and/or 'correspondent.'"

> Digital storytelling is an established community media tool which was developed at the University of California in the early 1990s (Lambert 2013; Lundby 2008; Robin 2008). Although various formats are currently available which range from simple story creations which include pictures and text to sophisticated multimedia production including images, recorded audio narration, music and video, the stories produced in the California tradition are short, focused 2–5-minute stories with visuals such as still pictures and the producer's own voice-over which centre on personal experiences or events. Both the personal point of v view and—often—emotional account are considered essential features of digital story telling. (Poeltzleitner, Penz and Maierhofer 2019: 408)

Digital storytelling has a presence in community activism and exists as a method beyond the "academy" and its practitioners share a recognized methodology. Participants work in groups with a facilitator, who uses artifacts and images to trigger conversations around specific topics. So, for the Veterans' Voices project, the veterans were asked to bring along photographs, certificates, and anything else that was from their time in armed service. In the accepted Story Center model, participants then work on a script that they go on to record as a short piece of audio (mp3). They, or the facilitators, upload this, along with their images and any photographs of the items they have brought along, to an editing software platform, in my case Premiere Pro. The audio and images are edited (by the participant ideally, but by the facilitator or technicians when this is not possible) to make a short, two- to four-minute digital story or "video," as many of the veterans continued to call them. These finished stories are shared via a project website and given to the participants either digitally or in DVD or USB format. Consent for public storage is given as part of the release and consent forms that participants sign, and they have the possibility of redacting their consent up to the point of publication.

The digital storytelling method allows for listening on two levels, first the listening that is done in the shaping of the story, when it is first "witnessed" as part of the scripting process and a second listening to the finalized "videos" stored on research websites, such as www.veteransvoicesglos.org where there is the potential for the stories to travel and to be listened to by large numbers of people. The stories are usually on a specific topic geared to the organization running them or the research project's aims. Mark Dunford, a key exponent of the method, argues that "Digital Storytelling is simply one means for people to acquire a media voice" (Dunford 2017: 316). Even if they might have been listened to before, by key workers, volunteers, NGO workers, this might be the very first time that they have used everyday media (their phones) to activate that listening, and for their stories to be packaged in such a way, presented on such a platform, and listened to by people they have never met.

Chazan and MacNab (2018) detail the power dynamics in their "intentionally feminist and intergenerational digital storytelling methodology" (2018). This emerged from a digital storytelling workshop that centered on "gathering activists and academics across four generations to share and record their activist histories" (2018). This was organized by "Aging Activisms" and the Ageing Communication Technologies (ACT), a research group led by Dr. Kim Sawchuk at Concordia University, Montreal. Their approach opened up "poignant moments of connection across generations" and had transformative potential. I like the approach they take, which is focused more on process than product, that is, they are more interested in what happens during the digital storytelling event than the finished mp3s that come out of it. They focus on the power dynamics that operate and fluctuate in order to rethink the researcher/participant dynamic. Their focus on process is mine too. Because the process of engineering shared listening and telling spaces carved out areas where all the project participants, I worked with could enter. The end result, the "video," was not for public broadcast, it was not going to be judged on its production or editing; it was there to foreground a voice, some of which, as Rick, a Gloucestershire veteran observed, had never been "shared" before (exit interview November 24, 2021).

These different projects presented challenges in terms of adaptation tweaks. For example, digital storytelling methodology requires that participants sit in a script circle, and work on their narratives with the guidance of a facilitator. What to do when a participant, as happened with my work with veterans, has written their script, in capital letters, on two sides of a scrap of paper and refuses to change it? How to do a script circle when there were only two participants and a key worker, when the person being recorded is a farmer in their "best" room? The logistics of the projects and participants determined that the method be adapted. It was more important at that point to listen to that person, the project, and its funders than to insist on a method that would not work for them. They wanted to tell their stories their way.

On both the Veterans' Voices and Erasmus+ funded "Mapping the Music of Migration" projects, I worked with an audio expert and freelance BBC documentary producer, Julia Hayball, who stressed to all our participants that what was important was how their story was told. Coming from radio and journalism, she used the KISS model, which means "keep it short and simple," something we had used to good effect in a previous project called "My Story," which used digital storytelling in schools in Austria, Hungary, Italy, and the UK. This did not work with veterans who wanted every detail that was important to them to be in their final story, some of which were over ten minutes long. Further adaptations were needed in the NetZero Project where my absence due to Covid in April 2022 meant that the farming participants were guided in their story recording by team members with no experience. Instructions were given over Zoom and five stories were recorded, two of which were played at the face-to-face climathons in Cumbria and Cornwall in May 2022. One of them was about using different methods to try to increase soil productivity, and the other was the story about surviving in the Cumbrian Hills which included that sentence we hadn't expected, "I got ill and my wife left me." Again, a jump, an emotional side punch. The unexpected. Showing these stories enabled these voices to be heard, as the farmers were too busy to attend them in person, bringing them into contact with the land agents, councilors, and agricultural partners who did attend. I made one, very important adaptation on the MaMuMi project, and switched out image for sound to be used as the trigger object. This is covered in Chapter 5 where I explain what the move from co-looking to co-listening did as the push to propel a story.

Connected Listening and Image

Digital storytelling projects traditionally rely on objects and images. They kickstart the stories. They are the artifacts upon and around which narratives interweave. I asked that veterans bring along old photographs and memorabilia when I met them for the first time at the Soldiers Museum in Gloucester on

February 27, 2019, and so they brought along meticulously curated photo albums. They were all notated, with names and dates. I wanted farmers and stakeholders to dig out photographs of their farms and local areas before I met them in May 2022 and when I did, we sat down in a sitting room and looked through the one family album they had. There were no dates or names to the photos here, the farmer had to try to remember who and what they were. The four participants on the GEM project I worked with across Gloucestershire were younger, in their twenties, thirties, and forties, and shared their photos with me on their phones. Across all those moments, we were involved in listening to images and I want to work out what that really means in practice.

Campt (2017) offers us a way into thinking about how certain images can speak, when their owners have had no voice; she talks about being alive to the "sensorial register of the image" (2017: 41). For her too, listening is "not just about hearing, but an attunement to different levels of photographic audibility" (2017: 41). There were occasions, in practice, when photographs were shared, where the importance of the photograph and its meaning in an individual's life was clear; it had historical weight. These photographs usually had a specific smell, the veterans' photos in particular had a specific weight and texture to them. Their surfaces were by turns, matt or shiny and they had worn edges. Many of them had historical weight to them, like a 1960s vinyl record. Like old records, they were "material artefacts [. . .] grounded in place, as well as in particular sociotechnical and affective-sensory formations" (Bartmanski and Woodward 2015: xi, in De la Fuente 2019: 563). They had a textural weight to them, their own "grain," a materiality that has affective potential in its innate sensory-ness (De la Fuente 2019); some of them were worn and torn, and in addition, there were many that had a particular smell that spoke of age and where it had been stored for more than fifty years. The photos were from the mid-1960s, 1980s and some from the 1950s. They were mostly in black and white. They were precious and venerated; part of the fabric of the lives I was asking about. They mattered. On top of this, we have to remember that they were part of a process that was different to how they might normally be encountered. This was no leisurely flipping through an album, on their own or with people close to them. The change of place (outside the home) and reason (a "project") changed the timbre of the telling. Veterans at the initial event on February 27, 2019, were looking at these photos and trying to ground a narrative with them for someone who was asking them for a story about their time in the forces. The people being helped via the GEM project were showing me, a relative stranger, photos of their achievements, farmers were having to take a day off to make tea for a stranger with a microphone.

Across all these listening projects, participants had to describe those images to me, and to others who might be listening in the initial meetings. These "reminiscence bubbles," where stories came up out of people and were let out, voiced, are where images became listened to as they were shaped by narrative demands and explanation. In these moments, photographic images were part of the discursive "memory work" that was going on (Kuhn 2007: 284), where iterations and re-iterations, sharing and confirmation, denial and forgetting, were all part of the process of telling and making meaning.

A lot of the listening happened in company, it happened together. These were social events, engineered yes, for research, but social all the same. There was a "hubbub" about them, they were noisy. Among the noise of all the conversations about the project, and general chatter, the stories that came to be produced began to emerge. In the Veterans' Voices, and GEM, projects in particular, there was a shared focus on the object or image enabled by a series of meetings and workshops. Spending time over a long period very clearly engendered trust and enabled convivial listening and sharing. For the MaMuMi project, the listeners were mainly team members and NGOs who had been working with the migrants and knew them well. The NGO listeners had had to participate in a specific Song Story Workshop that outlined the hows, whys, and wheres of Song Story collection (which is explained in detail in Chapter 5) and why we were doing it,

Figure 2.1 Veterans' Voices Project. ©Veterans' Voices http://veteransvoicesglos.co.uk.

so even when (as in Greece), extended team members conducted focus groups with migrants in a camp in northern Greece, there was a sense of security in terms of what the outcome should be. Across all of the listening points there was a focus across this visual or sonic third space. This space afforded multivalent connections that could spin off into new stories, as happened in the first meeting, where two veterans and a carer talked about the phrase books they had in Aden, which all soldiers kept on them in case of being captured (Figure 2.1).

Listening as Double Witness

Listening to people telling their stories produces a double witnessing. I was witness to themselves in their narratives, just as they were witness to their own pasts (recent or long gone) that they narrated via the memory prompts. Listening is witnessing and witnessing is about "being there." It is about sharing space and time and implies some sort of coexistence, however momentary that might be. It is a dialogic process, an exchange. It is important to note that I was not interested in the truth in any of the participants' stories, my role was to listen to them telling their stories in that moment on a subject. I was there, as were others present, to hear them, not to determine whether their stories had any truth. In this way, it might be said that I was mobilizing an understanding of witnessing that comes from Derrida's critique of Arendt (in Barbour 2011) although I was not using the witness of the law court or legal trial, more the therapeutic encounter minus the psychoanalysis (see Cohen, Grenier, and Jennings 2022). This entails a surrender to the integrity of the moment as "all testimony and all witnessing relies on an act of faith or a pact—whether it be spoken or unspoken, explicit or concealed—that entails sincerity" (2011: 631). For that moment, there needs to be an agreement from both parties that the dialogue is meaningful. Further meaning comes from the understanding that in all these projects, I was talking to people whose stories about what they had witnessed (in war especially) came across as personal testimonies and these "assume [. . .] a privileged relation to the truth because [they are] spoken by those affected" (Fernandes 2017: 5). Again, I was not interested in establishing the truth or otherwise of these stories. I was interested in what those listening moments produced in those spaces that were provided in the time available. I was trying to engineer spaces where people could be heard, whatever they wanted to say, however they wanted to say it.

A witness "does more than to give information about the past, however accurate or inaccurate. The witness 'from the inside' reveals the experiences and the good listener as to be with her as she is doing so" (Hirsch and Spitzer 2009: 162). This might be framed as a "politics of recognition" (Oliver 2015), where "affective and imaginative" lines of engagement might be established. A connection is formed through bearing witness, and if, as is the case in Holocaust testimony, "The listener to trauma comes to be a participant and a co-owner of the traumatic event: through his very listening, he comes to partially experience trauma in himself" (Felman and Laub 1992: 57). This "true hearing," which is a "listening for the emotional affective embodied truth of a witness's story" (Hirsch and Spitzer 2009: 162) is like a psychoanalytic reading, where the listening "places the listener at risk" (Hirsch and Spitzer 2009: 162).

Lines become blurred. Listening as a witness can be messy. It is, however, crucial and part of being human where that is an awareness of the self and others. Barwick's paper on "bearing witness" is written within a Kleinian psychoanalytic framework and his definition of witnessing foregrounds it as a "civilizing process characterized by the capacity to attend to an other's point of view, to discriminate and, where appropriate, integrate the value of that which is "other" while maintaining a unique sense of individuality in the process of doing so (James 1994)" (Barwick 2004: 133).

Connected listening is a mix of Dreher's listening across and witnessing that acknowledges the body and the ear and the messiness comes from the body that finds itself in these new formations. I have worked on storytelling projects where my background in feminist cultural studies, aging and music have impacted on how I approach and interpret the world as I move through it. My feminism is driven by a resolve to be treated equally and be heard, and I hope to have extended this remit to listen to the people in care, in old age, and in transit whose stories follow in the next three chapters.

Grounded Feminist Listening

Yes, this research is feminist, but none of the projects have been about gender. That may seem a little odd at first, but the emphasis here is on the "doing." They are all linked by an approach that I want to call Grounded Feminist Listening methodology. My research is formed by my own feminism and the needs of the research projects I took on; it is a compromise between the two, shifting and negotiating. It borrows from Charmaz's (2014) use of the term "grounded" but does so lightly, as I do not use memos or focused coding. I am very much shaped by "feminist theory, politics, and ethics and [the research I have undertaken is] grounded in women's [namely my own] experience" (Ramazanoglu and Holland 2002: 16) and I cannot exclude myself from the research process. I would like to think that the work outlined here emerges from my own "ethical and political obligation[s]" (Berlant 2011: 14) which focuses on listening to the voices of those who, for a variety of reasons, have not told their stories before, do not normally get asked to tell their stories, or do so in a different way. A long-time member of the AHRC-funded network of feminist cultural studies scholars "Women, Aging and Media," my work has been shaped by the desire to represent and empower older women and to research into intergenerational feminisms.[3] It is what I have written about in relation to popular music (Gardner and Jennings 2020, Gardner 2020). This feminist approach has not been confined to conceptual work. It is fundamental to the applied research under discussion in this book. Feminism is written into my body and cannot be taken out of it. I am, as Merleau-Ponty argued, "in the world through [my]body" (1964: 206 in Back 2007: 77). It is a lens through which I experience, see, listen to, write, and do everything, and as I age, it becomes even more pronounced. My body, from a Foucauldian perspective, is political, it is "involved in a political field" (Foucault 1994: 375). I have become more feminist in direct correlation to the increasing invisibility I experience in the public sphere, the more my body moves away from dominant youthful ideals, the continuing, weary reality of being under-represented in my professional life, of having to shout louder: of being a "difficult" woman (Brabazon 2002). This is the shape of me. And this "me" is marked through context, conflict, and conversation, in dialogue with expectations, fighting barriers, learning. In 2015 I spelled out why this was important:

> The constitution of the subject involved in the research process has been considered a key component of feminist research, where the personal has been deemed essential to the political and where the subject's investment in the enquiry has been accepted and reflected upon (Gray 1997: 87; Brunsdon 1997; Walkerdine 1997a; Walkerdine, Lucey, and Melody 2002; Burns and Lafrance 2002). (Gardner 2015: 54)

This foundational feminist thinking, which is, at its core, a locational approach, has anchored a burgeoning feminist autoethnography as a research methodology

---

[3] (https://uniofglos.blog/wam/manifesto/)

(Grist and Jennings 2020; Cohen, Grenier, and Jennings 2022). My approach is not autoethnographic although it is tempered by personal experience. This is because I am slightly uneasy about what Ratcliffe has called the risk of "lapsing into a narcissistic confessional solipsism" (1999: 213). I do, however, continue to reject, as Haraway did over twenty years ago, the "god-trick" of positivist epistemology (Haraway 1991c: 191–6 in Lykke 2010: 4) and welcome voices from researchers mapping themselves into academic methodological canons (Jennings 2020) and validating their experiences and research journeys. In sum, the listening that I have done and continue to do across a wide variety of topic areas and geographical locations is considered through an understanding of my role within the connections along which those listening's are enacted, and how, as a feminist in their late 1950s, I have listened to the erased and the co-opted, people whose lives have been subsumed under category titles ("migrants" for example, "veterans" another). Listening has been a way of breathing life and story into a person who has been categorized. It is about inclusion (Bickford 1996).

This feminist qualitative connected listening methodology acknowledges both the body doing the research and where the body sits, or where it speaks from. So, Susan Bickford using Arendt's work to argue about the importance of listening to democracy notes how "*what* we are (socially defined categories of race, class, gender and so on) affects *who* we are (our appearance in the public realm). Implicit here is an agreement with Arendt that others' perceptions of us affect how we can be present in the political realm" (italics in original) (Bickford 1996: 96). For whom "listening is understood as a shared responsibility to maintain connection and engagement, rather than mere individual engagement" (Dreher 2009: 449). Listening is about political and cultural vitality, about the entanglements of a listening democracy.

These entanglements illustrate how complex the vectors of communication can be across different bodies in different places. My (white, academic, British, middle-aged, middle-classed) body figures within what Lykke calls the "politics of location" (2010), a rewording of Haraway's "situated knowledges" (1991) that shifts the emphasis from who is doing the research to where they are doing it. All listening research approaches should consider the places where the listening takes place as those meeting spaces are the zones of interaction that produce the stories that become the research that is "output"; the short films or audio that are sent around the research team and its funders, and back, as USB sticks, Mp4s or DVDs, to the storytellers. As far as possible, all the listening I did across these projects was done in places known to the participants, and this approach formed the basis of the MaMuMi project as well, whereby research team members, who were NGOs, spoke with the migrants and refugees whom they had established working relationships with, in cafes and online. Back in 2009, I went to the synagogue to listen to the members of the Cheltenham Hebrew Community, and from there, to their homes. I traveled to the Soldiers Museum Gloucester, on February 28, 2019, to first meet veterans to introduce them to the idea of telling their stories. I was uncomfortable in this place, having been a CND member in my youth, but I tried to listen to what the building was saying to me, and how comfortable the veterans were there. Later, the meetings took place at the university's edit suites, in rooms full of high-tech computers and editing software, and one of the veterans, in his early eighties, later told me that he had never been to a university before in his life. For him, my work environment was his day out. An academic has privileges and power that need to be recognized and factored in, worked with, mitigated against if needed. There is no neutrality, and a feminist approach to research tries to reflect on that. I am also aware of the privilege that my whiteness brings, in that it has largely been unmarked, as Dyer so eloquently argued (1997). This means I am part of a culture of whiteness built into the university buildings where I work. Geographies of place need to be factored into the matrices of power that are present at all points of the research activity. These translate across in the online spaces where some of the research listening was done, as the Covid-19 pandemic began to hit.

Being involved in a digital storytelling project, finding your photos, having your story recorded is all about making meaning. In a review of autobiographical memories of veterans, Islam argues that veterans' storytelling via autobiography is a "meaning making process" (2021: 234). What is made in this process, I think, is determined by that process, which in turn is colored by the power dynamics of the place and people involved at the time, making meaning time and space determined. Architectures of place, project, and personalities all filter the final stories that emerge because "Listeners are . . . produced through the different mechanisms and manifestations of power by way of the institution" (Stead 2019: 180). As Voegelin writes, "Listening is never separate from the social relationships that build the fleeting circumstances of hearing" (2014: 1) and if the research method is based on listening, then we need to acknowledge these relationships. When working with the veterans, I was asking what it might mean to listen to and with, age. There was a shift in language I used, and a kind of acceptance of their mode of address and politics, that illustrates very well the flexibility I adopted to work with the organization (Age UK Gloucestershire) and these individuals. As someone who had been present at anti-nuclear marches in the 1980s and a lifelong Labor Party supporter, I had to rethink my thoughts about "soldiers" and the "armed forces" and tried to present a neutrality that enabled the listening exchange because I was "listening across difference" (Dreher 2009). This happened too in a project called GEM, where I listened to people who were part of an initiative to enable those a long way from employment, to take positive steps forward toward work. These people had little in the way of formal educational qualifications, had been exposed to criminality or neglect and again, I encountered them, this time, not in a university room, but in a local café, where I provided coffee and cake. In these projects, I was a middle-class white woman listening to those from working-class and Black and minority ethnic backgrounds. I was an outsider. The same was true for the listening I did in Cumbria and Cornwall, two rural counties at the edge of England. I couldn't erase myself, I have had to recognize how my "self" and how it talks, where it comes from, why it is there, what it is doing and what it wants. I also need to understand the reception it might get in these very different spaces. In Cumbria I was not only an academic, someone from somewhere else, a woman, a woman with a microphone and headphones; I was a southerner. And I have no idea what order people I was meeting saw me in. These recognitions are all part of the listening compass.

Listening as Care

Listening is aligned to care, and care has been configured as feminine. Why else would we have the Male Nurse? Only recently is care being relocated from its gendered location, and this depends on domestic politics and cultures (Sevenhuijsen 2003: 181) that are constantly reconfiguring. Listening is core to the therapeutic encounter; we all know about the "talking cure" that implies the authoritative listener, whereby experience moves into language, is aired, processed, in the listening therapeutic space. Listening to someone or something is caring for them or it. This relationship is not one of passivity although it has been conceived of as such. Being listened to has given me power, over my own past and its traumas and from there, to understand why I have conducted research that has cast me as listener. However, care is devalued, and the listening mode, aligned to caregiving, has suffered the same treatment.

Feminist work on care notes this well (Grist and Jennings 2020; Cancian and Oliker 2000).

> Women do most of the paid and unpaid caregiving, and caring feelings and actions are viewed as naturally associated with women. People tend to see caring as part of women's biological makeup or as a fundamental personality trait that corresponds to women's reproductive role. They do not view caregiving as skilled work that is learned through practice and shaped by cultural values and economic incentives. Moreover, activities are seen as caring because women do them. (Cancian and Oliker 2000: 3)

In care, listening is a "moral attribute" that is "developed through practice" (Robinson 2011: 856). Moreover, a feminist ethics of care (as argued for by Robinson) highlights how dependency and vulnerability are not "conditions to be overcome, but rather ways of being for normal human subjects" (2011: 845). The Veterans' Voices project and the Auschwitz Museum's Twitter feed were projects where individuals' "ways of being" were allied to their "making sense of" their pasts, which Gorton and Garde-Hansen, in their research on British Television and Memory (2019) argue, "has a connection to care and community that may get missed by critiques of structural change or political ideologies" (Gorton and Garde-Hansen 2019: 2). Again, this is an approach which, like LaBelle's, Western's and older cultural studies approaches (see Fiske 2011), sees the potential in small acts of consultation around pleasure, entertainment and lived lives and their associated failures and vulnerabilities. Vulnerability does seem to come through in a number of the research project listening situations myself and other research team members found ourselves in, where either the person talking, me listening or others listening, were mindful of a kind of frankness and exposure in proceedings that was unexpected. New connections had been forged.

I return to Roshanak Kheshti's, work on World Music, desire, and listening for what she has to say on the matrices of gender, care, and attention that are caught up in certain histories of listening. Similar to feminist care ethics scholars but working on the history of the American recording industry's use of "Other" music, she argues that musical listening "came to be constructed as a feminized practice through its increasing domestication by the gramophone industry and the ascension of the bourgeois woman" (2014: 3). Music moved into the domestic sphere and so became allied to the feminine, whose space that was. I am not convinced this ownership was maintained however, as many feminist media researchers have noted how control over the technologies in the house has been largely exercised by men (see also Gray etc.). She does say, however, that listening itself was to do with "pathways of connectivity" and "structure[s] of feeling" (Kheshti 2014: 38). Again, the stress on "feeling" brings us back to listening as feminized and in a paper on rhetorical listening, Ratcliffe notes how "Tannen claims that in our culture speaking is gendered as masculine and valued positively in a public forum while listening is gendered as feminine and valued negatively" (1999: 200). This is why the groom and the best man, and the father of the bride make speeches at "traditional" heteronormative weddings. The bride listens, silently looking on. Tannen goes on to argue how listening is further split along a binary gendered line whereby men listen competitively and women supportively so that "listening subordinates not only women to men but listening to speaking" (Ratcliffe 1999: 200). We are back to the stubborn seeing/listening binary that I noted in the last chapter; the supposed dominance of the visual which is not as inconveniently "close" as listening. Indeed, listening provides some kind of "contact zone" (Kheshti 2014: 22), forcing us into a relationship with someone, something, or somewhere else. It is about being with either co-temporally, co-spatially, or cross-temporally and non-spatially. Because of this it is not just able to be about cross-cultural communication, and cross-temporal communication. I have listened across class, age, and place. I have listened across time. This listening is stitched together by stories, told via digital storytelling, thought about through the lenses of care, politics, and proximity.

There are two considerations I want to cover before concluding. The first is about data and storage. Where do these voices, these stories end up? I asked at the start of this chapter how voices might "turn into" research data and what that means. It is easy to locate that voice when digital stories are stored on websites for public use. Age UK Gloucestershire wanted there to be a storage point, or archive of the Veterans' Voices, and on the site, there is an explanation of the project, its aims and each story is housed under a tab with the veteran's name on it. It is also neat, a kind of closure, when participants get their stories back on a USB stick and then go on to show them to others, as happened on the GEM project, or when they are shown to the funder as evidence of where policies are failing, as in the British Academy funded NetZero Project. But I think we need to consider what happens to voices when they are divorced from their bodies. The MaMuMi project developed an app that mapped voices onto a digital globe, pinning them to a place where they had traveled from and ending where their stories had been recorded and where they currently lived. I had wanted to make some kind of "family tree," some lines of inheritance that subverted both the

Google Map (so we had no borders on our globe) and questioned place and positioning. The second question refers back to Solnit's words at the start of the chapter. Why is this working taking place now?

It seems that there has been an awareness across some of the funders (Age UK specifically) that it is important to "capture" stories from people in their seventies, eighties, and nineties so that their experiences are remembered. European-funded projects have been keen to fund projects that foster inclusivity into the European understandings of citizenship and identity. UK National Lottery funded schemes that used digital stories as impact measurement, where impact was around self-esteem and assessing the validity and use of extensive support programs (GEM). They are all linked by some kind of urge to incorporate, and listening has been core to this.

The politics of funding, institutional will, and charity status have all influenced the choice and directions of these projects. I think that there were different things at stake in them. The first of these was not *what* was at stake, but *for whom*. The Age UK project and MaMuMi were the two projects that arguably produced the most resources, purely because they were funded for two years and this allowed for a development of connections across the teams and the people we were working with, veterans and migrants. The veterans often said that they were involved so that they, and their time in Aden, Korea, and Bosnia, would be remembered. Their motives were about memorials. For the NGO workers across Europe, MaMuMi offered a chance to listen to migrants about their emotional lives, not their legal, financial, health or housing worries and what is at stake for the team, was for migrants and refugees to have a platform where they could tell stories that were not circumscribed by their current identifying noun "migrant." All of the people who have been part of these listening projects had not been listened to before. But the EU, Age UK, GEM, and the British Academy all green lit projects that had "earwitness[ing]" (Canetti 1979: 43) at their center. They were all connected by funding that allowed their voices to be listened to, to understand how policies effect and affect them.

Conclusion

In the third decade of this, the twenty-first century, there are moves within the academic world to pay attention and listen to voices who have not been heard.[4]

---

[4] https://www.meccsa.org.uk/events/conference/

My work is part of this. In particular it is about listening to the ordinary, and to the singular/individual. It is about taking the time to listen and to mark, to note and to maybe hear a story that offers a riposte to those that circulate around "bigger" categories like "veteran," "migrant," "Holocaust victim." My methods are simple, but considered within the debates around listening, care, and witness and, although not in the therapeutic tradition, benefit from the space that the therapeutic moment enables. Connected listening is an adaptable and simple approach that can be used across different disciplinary arenas when we need to listen. It just requires that shift of focus, to stand up for the importance of listening. It takes time, and it can be painful; it can cause discomfort and disturb; the witnessing can make you jump at times, can force you into a different affective space. Connected listening requires the listener to be fully aware of what they bring to the listening and to be prepared to silence themselves, just for those moments. It asks that they be aware of their positionality and the multiple lines of connection coming from their own body, their past, their politics in that moment of connection to the lines coming from the voice(s) they are listening to. It happens on and offline. It is grounded, empathetic, qualitative, and affective. It is an approach which asks you to lean on listening a bit more than you do already perhaps, to allow time for it to emerge, and importantly, to understand how not doing it can be part of ongoing processes of containment or denial, of stereotyping and silencing. It is listening against erasure.

3

Listening Across Age(s)

The past is present as story. Often, this story is well rehearsed and well known. It rattles around the hallways of shared "histories," becoming what we understand to be "the past." These hallways might be personal, familial, national, or international, and the stories take all manner of guises, through photo albums, film, textbooks, television, and educational curricula. The echoes of these stories reverberate across these corridors, bounce off the walls, and wear themselves into the fabric of an understanding of what "has been." This chapter argues that listening, as a method of opening up, connecting to and across, has the potential to tweak those well-known stories. Listening is the interface that allows access to these places, which in this chapter are sites of conflict and loss, adventure, and trauma.

The chapter focuses on two case studies. The first is a two-year digital story project called Veterans' Voices, which was funded by Age UK Gloucestershire between September 2018 and June 2020 as part of a national scheme called "Joining Forces," which was an "Armed Forces Covenant Fund Trust-supported project to support the quality of life and wellbeing of veterans (anyone who has completed at least 1 day of military service) born before 1950 and their family and carers".[1] The second, is a critical analysis of my own two-year (and ongoing) Qualitative Online Listening engagement with the Auschwitz Museum's daily Twitter feed, Auschwitz Memorial@AuschwitzMuseum. I was working on them in parallel and started to think they might be linked in some way. They both tell individual stories of experiences of wars of the twentieth century. Veterans' Voices set out to be a conventional digital storytelling endeavor, whereas the Auschwitz Museum posts one or more photographs of people born on the day of the post and who were murdered there or in other camps, such as Theresienstadt. It also posts short threads on information related to the Holocaust. The two are marked by a concurrent mundanity and exceptionalism, that is, a juxtaposition of the

[1] https://www.ageuk.org.uk/information-advice/joining-forces/what-is-joining-forces/.

small and singular with the extraordinary. They both tell familiar narratives, of achievements, jobs, and family, which are positioned within the broader landscapes of the Korean War (1950–3), the British war in Aden (1963–7), the Bosnian War (1992–5), and the Holocaust (1933–45). The stories are small ones set within "big" arenas. Some are determined by familiarity and trauma, of recognizable life events and loss. They have this doubled layering, a forcing of the familiar into the strata of stories of war that have reverberated through the corridors of (the British) national curriculum and broadcasting. This is because I would say that they are acts of historiographical resistance, or recalibration, forcing personal, micro stories into macro historical narratives. They "color in" the larger narratives, the dates from and to, with personal artifacts and story. Some of the narratives to have emerged particularly from @AuschwitzMuseum's posts have upset in both their content and format, and the affective resonance of images and narratives in both, offer cross-temporal connections across bodies and time. This is enabled by a process of rippling, which relies on thinking about time as non-linear, and echoing; hearing again in performative historiographies to stories whose singularities and sonorities counteract generality and silence. I was listening with age, deferring to the veterans, and being reminded of my experience of seeing the Holocaust being broadcast on television in the 1970s and 1980s, and I was listening across age(s), listening to the past as it popped into the present. Doing this type of listening is crucial for reconsidering instances of global violence and trauma that are entrenched into the stories that circle around us here in Europe and beyond.

Veterans' Voices Gloucestershire

From September 1, 2018, to March 23, 2020, when the Covid-19 pandemic hit, I worked with students from the University of Gloucestershire's Media School and Health and Social Care in association with Age UK Gloucestershire on a project called "Veterans' Voices." Part of a national MOD-funded project called "Joining Forces," its initial aim was to encourage intergenerational conversations for veterans in their sixties to eighties which it was hoped would boost self-esteem (Islam 2021: 222; Caddick 2018; Long et al. 2021) and generate opportunities for conversations and sociability for veterans who lived within the large county of Gloucestershire, UK. Age UK Gloucestershire was working with SSFA, an Armed Forces charity (the Soldiers, Sailors, Airmen and Families Association, a UK charity that provides lifelong support to serving men and women and veterans from the British Armed Forces), who in turn were working in partnership with the national "Joining Forces" initiative. The idea was that Health and Social Care students would work to generate meaningful relationships with the veterans in the first year, talking to them about their time in the armed forces and their life afterwards. They would then, in the second year, work alongside students from the Digital Media and Film Production courses in the Media School, to script, record, and edit digital stories, which would be shown at a special screening at the end of the project. A project's life course does not always go the way it was intended and due to staff illness and Covid-19, Veterans' Voices became something slightly different from its original intentions. The Health and Social Care lead left due to illness six months into the project and so there was less time spent on cultivating the relationships over the initial year. Despite that, a series of events were held that followed the lifetime of a digital storytelling process. Health and social care students and their lecturer visited the Gloucester Farmers' Club on October 24, 2018, where a lunch was being held to work with veterans and brainstorm ideas, reactions, feelings associated with being in the forces and re-joining "civvy street."

On February 27, 2019, I went to the Soldiers of Gloucestershire Museum at Gloucester Docks to meet the veterans for the first time and to start discussing what they would use as artifacts, or as Helen Atkinson, the Age UK Gloucestershire lead for the project called them "items of memorabilia" and photographs that veterans had collected since the initial lunch. It was the first time I had met them, and the event was filmed by staff from National Age UK as part of the Joining Forces scheme. Helen is on film (on the Age UK website[2]) saying that the project aimed to "leav[e] a lasting legacy for younger people and for their own self-esteem. I think it's really good. They [the veterans] can leave that message behind in a digital way." The veterans had brought along photographs, certificates, and other "valued" items (Kuhn 2010: 303; Long 2021: 899), including photograph albums, certificates of discharge, army pamphlets (Arctic Survival, Jungle Survival, and Sea Survival), and an RAF English/Arabic phrase book for use if captured in Aden. It included basic phrases such as "Hello," "I am thirsty," and "I am sick and cannot walk" with the final phrase, at the bottom of the page, being: "Take me to the Political Officer and the Government will give you a large reward." Eight veterans turned up to the event, six of whom

---

[2] https://www.ageuk.org.uk/information-advice/joining-forces/veterans-and-digital-storytelling/

stayed with the project to the end, completing their stories. One of them, Emlyn, was the SSFA coordinator and he had brought along a flag that he had taken when he was in Aden. The navy, green, and black horizontal striped flag had a crescent moon and star in the middle, and blood stains at one edge. Emlyn was proud to show this flag, and it was filmed by the crew, but it was not something he talked about when it came to telling his story, and no one asked him how the blood got there or whose it was. The question went unasked. That absence was, I felt in retrospect, hugely telling of the community of remembering that was being activated in the project, a dynamic of possibilities of recollection, what was the ebb and flow of the permissible. Emlyn did not bring the flag along to further meetings, and it was not discussed in his story.

Emlyn, like two others at the event, was a veteran of Aden. He was a member of the Gloucestershire Branch of the Aden Veterans Association, along with Rick and Chris, who were the chairman and vice chairman, respectively. Other veterans at this initial meeting included Tommy, a veteran of the Korean War, who was an exuberant and humorous member of the group; hard of hearing and visually impaired, and hugely proud of being part of the armed services. He attended memorial services all over the country and was a "loud" presence. Willy (Alfred) and Alan had both done their British National Service, a system of peacetime conscription lasting from 1947 to 1960 whereby men between eighteen and thirty years old were required to work in the armed services for a period of eighteen months.

The third meeting took place in May 2019 at the University of Gloucestershire, where myself and Julia Hayball, a long-time collaborator and freelance BBC audio producer, introduced the veterans to the mechanics of the digital storytelling process. At this point, it was becoming clear what the group dynamics were and what kind of stories we might be getting. Here it was that we discussed scripting, recording, and editing, and showed two of the veterans around the facilities we had there. Rick had become a media production lecturer after buying his way out of the army, and Willy had been a projectionist in the army, and had later worked with BBC Bristol Natural History department as a camera operator and they wanted to check the "tech" out. We used a similar presentation via PowerPoint that we had used on an EU Erasmus+ funded digital storytelling project and found that the veterans stopped listening after around fifteen minutes. As lecturers, this was novel. As researchers, we had to adapt, and this was the first of many adaptations that we made in response to the people we were working with. They were just not that interested in how to do a digital story;

they were with friends and colleagues whom they usually met at lunches and what they termed "dos," social events with refreshments at community centers. They had collected their artifacts, some of them knew what they wanted to say. And that was that. This became clearer at the next event in July 2019, which was an audio recording session held again at the university. Alan, who came from Stroud in Gloucestershire, and prior to being in National Service, had not ventured further than Gloucester, had his script with him. It was written in uppercase on both sides of two sheets of blank A4 paper. He had taken on board the need to write a script at the May session, and he never deviated from it. No edits for him, no further explanations, no further tweaks gained from discussing his script with others. This was it. Again, this is not what is supposed to happen in the digital storytelling process, where scripts are iterated and reiterated in the round Alan had his story and he was not going to be budged. We wanted him to start with the sentence he had on the top of page 3 (Figure 3.1) as it was a great intro; short and punchy. But it was not to be.

The July event, held in a very hot un-airconditioned attic room in Age UK Gloucestershire's offices on the 22nd was one where there was slight disquiet. Alan had written his story; Chris and Rick, both Aden veterans, had not started theirs; Brian, a veteran padre from the Bosnian War, was in attendance for the first time as was another veteran, Dennis, who turned up just for this event. Tommy, the Korean War veteran, was there, as he was at all events, only this time, he was being filmed by a two-man Korean TV crew. None of these factors

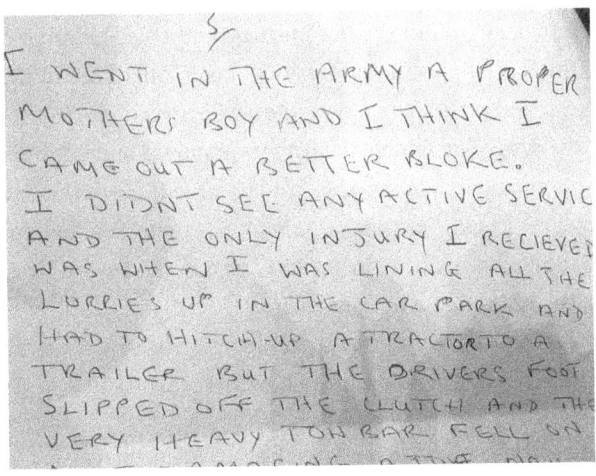

Figure 3.1 Page 3 of Alan's story. ©Veterans' Voices http://veteransvoicesglos.co.uk.

contributed to a successful event in terms of scripting, although Rick brought along his certificate of discharge from the army dated August 8, 1969, that had cost £150, which did make it into his story. What became apparent was that the participants were at different stages of the digital storytelling process. There were some veterans for whom the idea of editing their story was not going to happen (Alan), and for some, the story was still dependent on their finding photos and objects that was taking time. It was becoming clear too, that Tommy, who was hard of hearing and sight, came into this story at different points each time he tried to start. These points were never narrated in chronological order; how he joined up, how he ended up in Korea, how he walked 500 miles as a prisoner of war, how he escaped, twice, from a POW camp. He evaded any attempt to help him script his story, preferring to go straight to an audio recording, of which we made at least four or five over the next two meetings. The photos he brought along in October could be put, more or less, in chronological order, but that is not how he accessed his past and narrated it.

At the October event, held at the university, Sally joined the project. She too was an Aden veteran and the only woman to be involved in the digital storytelling. Like Alan, she had a firm idea of her script, although she had not written it down. Instead, as we had suggested, she worked on a narrative that followed a series of images. I have used her story to illustrate the effectiveness of the project and it is worth telling it here. She joined the WRAF (Women's Royal Air Force) in 1962, first training as a wireless operator and was posted in Aden in 1964 aged eighteen years, after applying to go to Germany, Singapore, or Cyprus. After three months there, her commanding officer approached her having found out that she had a background in show business and asked her to produce a show for troops in the Radfan mountains. She was taken off admin duties and found dancers, singers, and musicians out of the 10,000 men and 150 women based in Radfan at the time. The female dancers were nurses and administrators, and there was also a rock band, an ex-end-of-pier comedian who oversaw sketches, and Sally herself, who sang jazz with a quintet. She then took the show elsewhere in Aden and fulfilled a lifelong dream to entertain the troops (she cites *White Christmas* as an example). The entire story seems delightful and endearing, entertainment in adversity. But she then changes register when talking about visiting the wounded in Aden and notes how after this visit:

> worse was to come when, a few days later back in Aden, three of my mates were blown up in a jeep. They'd thrown a grenade in the back of the jeep on the way

out and they all died. And to go to the funeral in Silent Valley, which is just a desert, and watch the sand quietly across whilst the bugler played The Last Post brought the whole experience of Aden really and brought myself back down to earth to realize what the whole thing was about.

Hearing Sally record this on the day was unexpected. She had arrived at the meeting with her script fully worked out, her images in sync. Her middle-class Southern English accent exuded competence. She had a reassuring and authoritative presence. She did not make eye contact during or after the recording, and never referred to it again, preferring instead to discuss social arrangements. It might have been that this was the space she could use to sound out that horror, and for it to be listened to, as she let it go. As Cohen, Grenier, and Jennings (2022) note in their work on talking about the past, our work was not intentionally therapeutic, but something of that therapeutic exchange, or place, might have been afforded to Sally in her releasing that story. There was another layer of silence in that story; we don't know who "they" were, apart from the unidentified enemy, the reason the British were in Aden. Sally's story sat uncomfortably with me as I showed it to others, at conferences about "silenced voices" (IASPM 2022[3]; MeCCSa 2022[4]) and the weight of the unnamed started to be heard (see Figures 3.2 and 3.3).

Figure 3.2 Sally (right) and her dancers. ©Veterans' Voices http://veteransvoicesglos.co.uk.

[3] https://www.liverpool.ac.uk/music/events/iaspm/
[4] https://www.meccsa.org.uk/conference/

Figure 3.3 Sally Nash and the Del Turner Quintet. ©Veterans' Voices http://veteransvoicesglos.co.uk.

The next two events were held in edit suites in the University's Media school. Students from film production and digital media courses were paired up with veterans to help produce the final stories. Rick had some experience of using Premiere Pro, the video editing software, and really took to the process, as did Willy, who had worked as a cameraman for the BBC. His health prevented him from making the actual edits, but he was very clear on where they should be made, directing the student on when and where to edit. Him and Rick challenged "ageist stereotypes (e.g., the assumption that older people are technologically illiterate) [where] younger participants [are] receptacles and technical experts" (Chazan 2018: n.p.).

The veterans all returned the following year on February 18 to make the final tweaks to their stories. This involved getting all the images in the order that the veterans wanted, and some, who had brought along large numbers of images, were quite anxious about this. Chris in particular, whose story is the longest at twelve minutes, was not to be persuaded on cutting his story any shorter and the photos on screen had to mirror those in his photo album. He was displaying a resolute allegiance to narrative, and, as Kuhn wrote, the photos were not just props but lived inside this album which was "storied":

> Photographs operate as props and prompts in verbal performances of memory, but the collection of photographs that makes up a family album itself follows an "oral structure": An album is a classic example of a horizontal narrative shot through with lines of both epic and anecdotal dimension. (Langford 2001: 175 in Kuhn 2007: 285)

The plan was to put all the stories onto the website that Digital Media students had built and have a public screening in May that year at which the veterans would all receive a copy of their digital story on a DVD or USB stick. Of course, this did not happen. From March 23, 2020, all face-to-face and on-campus activity at the university was suspended while the series of lockdowns were implemented to try to contain the spread of the Covid-19 virus. Eighteen months later, on November 25, 2021, the screening took place at the university and all veterans bar Willy, who was poorly, attended, along with members of the armed forces charity SSFA and executives from Age UK Gloucestershire. I arranged for three journalism students to conduct interviews with the veterans on the day, asking them about what they had got out of the project and what, if any, impact it had had on them. Rick noted that the project enabled him to gain insights into people's lives that he had not heard before.

> Um, I think one of the things that happens in the forces, you don't always share stories with people. You sort of keep a lot of stuff to yourself, unless your friends already. So, what happened here is the stories of other people and some of them I know quite well and I didn't know anything about their backgrounds or where they'd been or where they served, what they did in n their days in the forces. That was one big thing. And the other thing was just getting together with veterans and telling stories about the times that were different for all of us but were enjoyable times. (Rick, interview November 25, 2019)

Sally too said something similar to Rick about not talking about certain experiences in the armed forces:

> I think watching this this afternoon, what you gained, we had no idea. We've all known each other from The Aden Veterans Association for 10 years, it's an association started in Stroud, but none of us knew about our time in the services. We just don't talk about it. So, this has been really interesting to see what the boys have been through, not me, but what the boys, you know, they've been through. It is a camaraderie that you don't get anywhere else. (Sally, interview November 25, 2019)

It appears, therefore, that the process of telling and listening to these stories moved historical events and individual experiences from the "unknown to [the] known" (Long 2021). In their work with veterans, Long makes the point that part of the impetus for doing this kind of work is to "mak[e] histories knowable" (Long 2021: 899). Some of the histories were less "noisy" in the corridors of national memory, and Long notes that:

> There was a trend for those who had not been deployed for combat and those who had served for 2 years in National Service to compare and situate themselves between older personnel who served in World Wars and younger personnel who served in the later Gulf Wars. By situating themselves between these periods, they explained that others had it harder suggesting that their experiences were not as interesting or important in comparison. (Long 2021: 899)

Alan and Willy told their tales of life in the armed forces as part of National Service. Alan had always wanted to be a driver, he says, "I was very keen to learn to drive and asked to join the Service Corps," which he did. He was enlisted in 1953 and stayed in service until 1957. "I didn't see any active service. The only injury I received was when I was lining up the lorries up in the car park and had to hitch a tractor to a trailer. The driver slipped his foot off the clutch and a very heavy tow bar fell on my foot." Alan wrote how when he joined the army he was a "proper Mothers Boy" but when he left the army, he was a "better bloke." Alan turned up to every event; when interviewed at the final screening, he said that he had never been to a university before. How important the project was for him became clearer halfway through, when his wife went into care, and he was alone in his house. Being a veteran and having access to their social events along with this Age UK Veterans' Voices project gave him a chance to be with others, to listen, and be listened to within a welcoming and supportive community with shared experiences and pasts.

Remembering in Place

The Veterans' Voices project offered an extension to that community for the specific purposes of collective and individual remembering. In her 2002 work, Annette Kuhn asks us to "Consider the photograph's currency in its context or contexts of reception. Who or what was the photograph made for? Who has it now, and where is it kept? Who saw it then, and who sees it now?" (Kuhn 2002: 8). At the first meeting in Gloucester Docks, photo albums were opened outside the home, with different people looking at them. Certificates of enlistment and discharge were passed around, there was laughter and recognition. All of these artifacts and images functioned as "occasions or communication, cross-cultural exchange, and cultural continuity . . . there is something distinctive about the discursive features of these image-based communications, the kinds of talk that accompany the viewings of family photographs and albums" (Kuhn 2007: 285–6). This "kind of talk" was as much about listening to others' experiences as much as telling their own, and in that listening was recognition and discovery. In her work on music and memory, De Nora talks about the "alchemy" of a shared "resonant" moment (2000: 67) and there is something in that which encapsulates the connected listening that happened across the project in its eight events. These were dialogic spaces where conversations built story promoted by the recourse to the material artifacts and photographs, and as Long argues, it was through sharing "materials" where discussions ensued that enabled "feelings of connection and belonging" (Long 2021: 892). There was, across the project a "communit[y] of remembering" (Kuhn 2010: 298 in Long 2021: 894) where, as Fivush notes, "One's own story is embedded in the stories of others in the past and in the present" (Fivush 2008: 55).

Veterans' Voices was driven by a methodology that prioritizes image and artifact, and I want to stress the tactility of those things. Some of the pieces of paper, the pamphlets, and the flag were over sixty years old. They had a patina to them; worn in places, creased, an aura. They were testaments to events, performative documents as well as representative images. And some of them smelt over sixty years old, with the musty smell of old film, old material. This materiality meant that the past "burst into the present" and wasn't "that far away" (Tanner 2021: 245). Tanner argues that "we carry the past, even the darker chapters, with us in the present, so it isn't totally gone forever" (Tanner 2021: 245). This was apparent in the grain of the voice of the veterans, the hesitations of Alan when recording, the reference to the "half-crown" that Willy earned, Tommy and Emlyn talking about the army as an escape from a life of drudgery, down the mines, and early marriage. From the initial conversations about the objects that they brought with them in February 2019, to the stories listened to in November 2021, age was etched into material and voice. Long's work on veteran storytelling talks about "the performativity of materials as repositories of memory and instruments of social performance" (2021: 896), and this is something that became noticeable as the project went on, and which I want to call "performative historiography," which is where the repeated, rehearsed, and delivered personal narrative becomes "the" story about that past, expected, and unrefuted.

Performative Historiography

The performances of memory that Kuhn talks about, the re-telling, the repetitions, the scripts, were part of a performative historiography whereby the veterans' stories became what was shared about their pasts. Long's work on veterans and the "materiality of reminiscence" is focused on making "unknown histories knowable," and they use the term "performative" in the following way: "For us performative potential refers to the possible effects of sharing materials with others to enhance the understanding of the person who was not there, making their histories more accessible/knowable" (Long 2021: 895). Their research indicates that the discussions and sharing of such materials (photos etc.) "helped them to make personal histories knowable" (Long 2021: 899) and this was the case in Veterans' Voices, where individual life experiences were "storied" and shared. Further work on social memory by Welzer goes some way to extending this idea of personal histories remembered in conversations, what he calls "conversational remembering" (Welzer 2010: 5). The key to this is that "Social memory exclusively exists between subjects and not within them; its form of existence consists of communication" (Welzer 2010: 5). These observations about veterans' social and conversational remembering foreground the role of exchange in the manufacturing of the past, but there is, I think from what I saw in Veterans' Voices, something else at play.

First, processes of selection and discarding were in force when the veterans were engaged in the process of telling and retelling their pasts. The narratives were clearly and expressly edited, at our request it must be said, and through the selection process that the veterans embarked on. So clearly, the stories were edited at every stage of their production because we, the project team, asked them to do so; in essence they were in a story space, where all of us were working with shared expectations of veteran life narratives. Second, the scripts of these military pasts were "frozen." Alan scripted his story in detail at the start of the project and did not change it throughout the whole project. His first version of his story of his time in National Service was the final version. Similarly, Rick, Chris, Sally, and Brian chose photographs and medals, discharge papers that they arranged in chronological order to tell the story of their entire time in the forces (Rick), in Aden (Chris), or of their role as a show producer and singer in Aden (Sally). Tommy was the oldest of the group and the last to make his recording and this was because he kept coming into his story of his time in the Korean War at different chronological points. His recording is one of the longest (11.6 minutes) and has Helen, the Age UK Gloucestershire manager asking him questions to prompt him into a particular story that she knew from having heard them over and over before. He had joined the army young on VE Day without his father's consent, had fought at the Battle of Imjin, been captured, and become a prisoner of war. He had escaped twice. He repeated details from those stories such as hearing a trigger being pulled behind him as he and another prisoner escaped. These fragments were the keys to his final recorded story and were performed by him at every event over the two years. He delighted in telling them to new people (including the three-man Korean film crew in July 2019) and they became an accepted version of his wartime past, that was then broadcast to others, face-to-face and then, via his recorded story, via video. His story contained lots of information on battle positions, which countries were fighting where and some of these were muddled. He was dealing with surfacing memories that were his knowledge "from the past . . . not necessarily knowledge about the past" (Margalit 2002: 14), but they were his memories, that we, as a research team, were trying to get him to form into a linear story. Tommy has been interviewed about his experiences,[5] and he can be seen on the Korean War Legacy site talking about landing at Pusan (in South Korea) and being taken prisoner by the Chinese. He must have told his stories many times.

These repetitions of the stories over the two years, in groups and then online, were part of a process of performative historiography, whereby the stories were rehearsed and repeated, becoming an accepted version within the group of participants and facilitators of their own and others' histories:

> recollections (les souvenirs)—what we retain in memory of our past experiences—are not just simple imprints; they are truly active selections and reconstructions of this past. Individual experiences, even of the most private, personal, and intimate nature, are the result of an ongoing dynamic social process; they are inscribed in a given, physical, sociohistorical environment, stored in memory and recollected through continuous interchanges with significant others or significant groups. (Apfelbaum 2010: 85)

Murdock's work on memory talks about the "veracity of the masks we each wear; the stories we tell about ourselves" (2003: 55). Some of those masks were made up more fully and reiterated (Tommy, Rick, Brian) and some of them were let drop (Sally). Maybe this storytelling space was somewhere that Sally felt more able to

---

[5] https://koreanwarlegacy.org/interviews/tommy-clough/

narrate things that had not been heard before, and Chris used the opportunity too to mourn his friend, the priest killed in a plane crash. It also became clear that there was an immediacy to many of these recalled events, and their affective impact on the present.

Rippling

Over the course of those two years on the project it became evident to me that time was operating in a way that others have noticed, and which exemplifies Barad's thinking on the subject. Scholars of memory studies often target the imprint of the past in the present, portraying the coexistence of different time layers in the same moment (e.g., Huyssen 2003), or the multidirectional temporal movements between past, present, and future (e.g., Rothberg 2009; Erll 2011; Hristova 2020: 778). I want to think about those moments as ripples, in that the past ripples into the present in the same way that ripples move across water. Just as ripples are created by the interaction of wind on water, or of an object displacing water, so the past ripples across the surface of the present. The main point from this analogy is to approach listening to the past as a process like thinking about how it ripples into the present. The past is on the surface of the present and all of us in the Veterans' Voices project were involved in listening to it, through these stories about Aden, Korea, Bosnia, and National Service. And just as the ripple is the result of that physical interaction, so listening is the interaction with the telling which produces that time ripple. It works in a similar way to Barad's argument about the relation between the past, present, and future, which is "not in a relation of linear unfolding but threaded through one another in a nonlinear enfolding of spacetimemattering, a topology that defies any suggestion of a smooth continuous manifold" (2010: 244). "Threaded through" gives us a kind of knot, a junction where different times interlace, or "percolate" (Tanner 2021: 53) with each other. I'd say that some of this interlacing becomes apparent through listening to the self as it narrates the past. At times that listening will "spook" (as Derrida's spectre haunts the present), where it's "time is out of joint, off its hinges, spooked" (2010: 243), at times it will be difficult to unravel the knot, and "different times [may] 'bleed through one another'" (Barad 2017: 68). This temporal relationship has been described with verbs that are about a lateral process (percolate, bleed) and fear (spook) and recall the descriptions of the invasive potential of sound and listening (Erlmann

2014). These move the relationship away from being a linear one to one which is messy, co-existent, and sometimes scary.

> The process of remembering the personal past is always already permeated by narrative as well . . . that exists the moment we try to make sense of the movement of experience. Indeed, it has been argued . . . that the process of living is itself permeated by narrative, that indeed to be human is to live in and through the fabric of narrative time." (Freeman 2010: 274)

Work on this project suggested to me that this fabric is, in parts, torn. Narrative still drives digital storytelling, it is dominant, expected as a shared and secure navigation system that delivers the past, as Kuhn also writes:

> It is impossible to overstate the significance of narrative in cultural memory—in the sense not just of the (continuously negotiated) contents of shared/collective memory stories, but also of the activity of recounting and telling memory-stories, in both private and public contexts—in other words, of performances of memory. (2010: 298)

But there are holes in it; there are gaps and rents in the narrative process and into these gaps comes kairotic time, the short-lived but exceptional time of intense experience (Gardner and Jennings 2020: 13). Sally's recollection of hearing "The Last Post" in Aden as her three friends were buried illustrates this, where her short remark "three of my mates were blown up in a jeep" acts as a kind of sonic punctum. It stands out and ripples into the present as a quietly delivered monstrosity "they all died." Other "big" events, about death, travel abroad, prison, and illness; figured in the veterans' stories; people lost (Chris's tale of a priest killed in a plane crash in Malaysia), Tommy's throw-away reference to "hearing a gun click behind us" or "a slow march of 500 miles." These had been moments of fear and hardship, of being almost starved. They are memorable because they were exceptional, they make good stories, they stand out. There is something about the relationship between the exceptional and the mundane that is key to this rippling which listening to one's own and others' past affords, and that digital storytelling enables. It is a lo-fi and seemingly mundane format, it's little more than a slide show at its core, just a few still photos and audio, but it allows for exceptional insight into people's lives, which is something that the veterans acknowledged in their exit interviews. "As Zygmunt Bauman pointed out, the most important things often remain unsaid" (Back 2007: 95). Maybe so, but the veterans used this opportunity to unlock some of their own important "unsaids," voicing some of the "spooks" of the past (Sally's loss, Willy's injury).

They also brought into being remembrances of wars that hover in the margins of British popular history, Aden, Bosnia, and Iraq. These were messy wars, complex and contentious, where the British legacy is more ignominious perhaps than the "big" World Wars of 1914–9 and 1939–45 that figure in national memory acts.

Upsetting Narratives

The veterans' stories were upsetting in form and in content. I don't mean "upsetting" to mean distressing, but in a circumvention of the expectations of veteran narratives. Some of them resisted chronological explanations, instead coming into an event from multiple time entry points and at multiple times. They were "disrupt[ing] the conventions of narrative forms" (Barad 2010: 244) and Tommy's in particular did this. The facts of the Second World War, VE Day, joining up and stages of the Korean War were all interwoven and difficult to entangle in his accounts.

In their oral history work with veterans, Islam notes that "The veterans' memories, examined through oral history interviews and memoirs, tended to follow the structure of a story, starting with a time they joined in military followed by war experiences in chronological order to the termination of the job" (Islam 2019: 220). This was the case for Alan, Rick, Chris, and Emlyn, who could map that chronology with clearly defined photographs and artifacts. Sally's story was more concise, moving quickly from her tale of joining up to concentrating on Aden and the shows she put on in there, before inserting a few sentences about the bomb that blew her three friends up. Tommy's narrative was chronologically incoherent, as he started off with the story of his escapes from a Korean Prisoner of War camp, before going back to talking about when he joined up. He upset our fine-tuned digital storytelling process, which was based on coherence and linearity. Working with him over the two years, however, we managed to record him a number of times and could edit in chronology in post-production. He seemed happy with this, but, possibly because of its verve or even heroism, the escapes and the 500-mile march across Korea were always going to feature more than stories about joining up, although he did this without his father's approval and was underage when he did become a soldier.

Some of the veterans' content upset larger histories of Aden in particular, and of Korea and Bosnia. Aden is not a war that is discussed in the UK as part of any history curriculum that I recall taking. So, what listening to these stories did was move this event from the "unknown to [the] known" (Long 2021). Two veterans, Brian and Emlyn, brought up, in interview, the topic of a broader politics of the wars that had been the "theaters" against which some of the project's stories were told. For Brian, learning about Aden had helped him to understand that the current situation in Yemen was with relation to the group, ISIS. And he mentioned that in conversation (but not in the published stories) he had noted some regret among Aden veterans, that they had "pulled out before their mission was complete." Aden might be termed a "forgotten" war (Back 2007: 95) but the Gloucestershire Aden Veterans Organization had this chance to reflect on their role within it. It was only Sally who mentioned death, Chris's story centered on where he had been and what he had seen, like Tommy, who had joined the army on VE Day in 1945 and said "I wanted travel, and adventure... off I went to Korea, even going on the boat was an adventure, you know, like a cruise. We'd never been further than the local holiday camp, you know." I began to wonder if there was an unwritten code in veterans where their experiences of loss, death, impairment, and battle were pushed to one side, or dealt with perfunctorily (Margalit 2002). For example, Tommy had a picture of himself and another soldier in a trench in Korea loading up some machine guns. He talked about it being the Battle of Imjin, the "final battle. We were completely surrounded. In fact, when we were captured... we were taken prisoner and walked five hundred miles north to a prison camp." Emlyn's reflection on the process of remembering contained quite strong emotions around his place in history and seemed to contain a defensiveness about why he had been in the armed forces, and how it was the fault of the generation above him. He said, in an interview at the final screening event:

> It's good to remember. It's good to tell your story because your generation don't have a clue what it was like when I was born. I was born into a war. Your generation thinks that we're ruining the climate. But my generation before me left me in the middle of a war, or the start of the war, which is worse than climate change. (Emlyn, interview November 25, 2019)

Willy's (Alfred's) story upset accepted narratives of army life in a humorous way. He concentrated on money, specifically on army wages, on the financial rewards of being a conscript and how he could earn more in the army than he could get in "civvy street." His tale of being called up in 1958 aged just seventeen, of going to Hampshire, to "Blackdown Camp" where he was declared medically unfit and being invited to run the camp cinema, as he had experience of being a film projectionist, was punctuated with the comment that "With £9 a week and £12 a week from the camp cinema, I was doing pretty well."

Listening with Age

Tommy's, Emlyn's, and Willy's stories exemplify something of what I want to say is "Listening with Age." This type of listening happens when two or more different generations listen to each other and there is an awareness that age is clearly marked in the narrative of the older person. This marking comes in the age of the mnemonic devices, it comes in the texture of them, and their status as coveted objects hitched to a particular time in the past. This marking comes too in the language of the older person's narrative, the reference to extinct monetary units ("half-crowns"), to "holiday camps," "jazz trios," and "end of pier showmen." It also comes in the meter of the story and its delivery; for those who had not written their scripts down, there are hesitations and self-corrections (of facts, "An RSM, no he wasn't an RSM, he was a Sargent but not an RSM").

Listening with age is also about adhering to convention and expectations. I am someone who was twenty to thirty years younger than the veterans, and the student post-production team and journalists were fifty to sixty years younger. During the recording of the stories and in their editing, I would say that we were performing respectful listening with age (Gardner and Jennings 2020) which was coupled with an acknowledgment that we were on a project funded by Age UK that was seeking to give veterans room to speak about their armed forces experience. Our remit was not to question their roles in Aden or Bosnia; our (at least my) politics or inquiries were not welcome in that space. When I recorded Brian's story about his time as a vicar in Bosnia, he referred to his time in "Rhodesia." What must a listener do who is trying to hear stories when those involve comments about battles, deaths, and colonial pasts that made me uncomfortable? As project lead, I was not tasked with agreeing with what emerged from these stories, nor to ally myself with the veterans' politics

We knew the type of listening we were required to perform at that moment, with those people, and for that project, it was empathetic rather than critical. The listening with age that Julia and I did with the veterans was also about having a willingness to suspend judgment in order to hear (Campbell 2008: 42). This does not mean that moral or ethical judgment is removed, but it is about opening out and sharing a space where stories can breathe. In his work on music and memory in Jewish Holocaust survivors, Toltz argues that "listening is immediacy: in listening, I create an attentive space where the musical memories of Holocaust survivors are allowed to resound" (2016: 195). The veterans listened to themselves and to each other and we listened to them. In doing so we created a multiplicity of spaces that now exists online, for that listening to be done. The next section is about a very different space.

Listening to @AuschwitzMuseum

I have been a silent listener to the Auschwitz Memorial account for over two years. Every day I have listened to stories that emanate from a series of online images. Every day, the Twitter account "@AuschwitzMuseum" posts photographs of individuals who were born on that day and were victims of the Holocaust. The account has 3.1 million followers at the time of writing, and posts twelve stories a day, to go out at hourly intervals. Each post includes a photo of a person accompanied by some facts about where they were born, who their family was, and what job they did if they were an adult. Most entries finish with the phrase "they did not survive." There are several photos each day: European Jews, Polish clergy, Roma and Sinti civilians, Soviet soldiers, and French criminals. I have been following the account for over two years and now, I start to recognize people from the previous year's post. They have the power to stop me in my tracks, to make me pause, to think back, and to be both sad and angry. I feel what Stewart calls, the "charged particularity of (the) objects, images, and events" (Stewart 2007: 21). She notes how "Encounters can happen anywhere" (Stewart 81) and so it is that by looking down at my phone as I scroll through Twitter, I see explosions of tragedy spliced with images of hope, love, and the everyday. These online short-burst stories and images make up a contemporary heritage culture that is arguably part of the affective economy (Ahmed 2004), reconfiguring dominant narratives of the past and providing emotional connections across time. Ahmed analyzes the discourses of emotion across an American far right-wing website and uses that as the platform from which to ask questions around "How ... emotions work to align some subjects with some others and against others? (2004: 117). She argues that emotions shape individuals and communities, "they create the very effect of the surfaces or boundaries of bodies and world" (2004: 117), to the extent that "In such affective economies, emotions do things, and they align individuals with communities—or bodily space with social space—through the very intensity of their attachments" (2004: 119). This intensity is at work in the @AuschwitzMuseum's fee, their format and content adding into that intensity which has the potential to puncture my every day, enfold me into someone else's horror, align me. Kathleen Stewart's work on ordinariness and affect looks at their relationship as follows: "potentiality is a thing immanent to fragments of sensory experience and dreams of presence. A layer, or layering to the ordinary, it engenders attachments or systems of investment in the unfolding of things" (Stewart 2007: 21). This word "Unfolding" appears again, a kind of wrapping into, an alignment. Today, May 2, 2022, I can scroll through Twitter and see posts like this: "12 May 1901. A Polish Jew, Moses Anger, was born in Radomysl, Wielki. A worker. In #Auschwitz from 24 February 1942. No. 25186. He perished in the camp on 3 March 1942." This text is accompanied by three photos of Moses in camp uniform. The first is his right profile, the second full face, and the third the left profile; pictures taken on his arrival in camp. The first picture has text on it "Pole 25186. J. KLAuschwitz." A few tweets above him (I follow around two thousand accounts) is this: "12 May 1911. A Pole, Antoni Tracichleb, was born in the village of Krupe. He was a clerk. In #Auschwitz from 3 November 1942. No. 72241 He perished in the camp on 17 January 1943." The same three photographs are there, this time Antoni is classified as "P," Pole. On the following morning, Friday, May 13, another series of images peppers my Twitter stream. One is about Suzanne Feldman, born on May 12, 1917. "A stenographer. During the war she lived in France. She was deported to #Auschwitz from Drancy on 2 September 1943. She did not survive." Underneath the text is a photo of Suzanne. She is facing the camera and her face is cupped in both her hands, elbows resting on what looks like a zinc café table. She is smiling. The photograph has a slight tear across it. This image stands out to me. Suzanne is so radiant. I recall seeing it last year, as I did on April 5, 2021, when I noticed a photograph of a young girl who looked like my mother: "5 April 1933 | A Belgian Jewish girl, Genevieve Annette Flora Levy, was born in Saint Gilles. She was deported to #Auschwitz and murdered in a gas chamber" (Figure 3.4).

This photo had over thirteen thousand likes. It clearly resonated with a large number of people, sounding out and across the online stream. The resonance here, the sounding out and seeping into, is also emblematic of the type of affinity work, which Twitter activates and relies on. It offers this small space, a moment where an emotional connection might take place, a jump, a reaction, a mode of "attending to things" (Stewart 2008: 73). A like, a dislike, a recommendation. A way of engaging people. And it is very effective.

Listening Across Age(s) 83

Figure 3.4 Genevieve Levy. @AuschwitzMuseum.

To discover more about this account, on April 2, 2022, I talked on Zoom to Pawel Sawicki, Press Officer at the Auschwitz-Birkenau State Museum. In answer to the question as to how long the museum had been doing the daily posting photographs on Twitter, he said that it had started around two years ago and it was an attempt at

> Different ways of linking the history and the story on daily basis and this is what we are basically trying to use the social media when we commemorate and educate about the history of Auschwitz to make people aware that you know except of those main anniversaries that public attention focuses on Auschwitz, like the liberation date or the first transport of Polish political prisoners. . . . Oh, there are some major dates, but it was also important for us to tell people the story on different days and something that was not possible several years ago. Now it is possible because more and more archives are available online, there are some photographs databases that had not been available before and when our archive made it available and we could get the IT was very easy for me to get the pictures of registered prisoners and then I started adding 1,2,3 faces and names in a day. However, it is of course important for us to present this very various context of this story about what's in the memory of Auschwitz, so we cannot

just concentrate on one group of prisoners. We need to show the diversity and therefore I started searching whether it will be possible to get photographs of Jewish victims and other victims and step by step I managed to kind of set up the right regulation so that every day there will be about 12 names so that every two hours there would be a tweet with a name with a person. And so, I've managed to do it . . . for the last almost two years. (Zoom interview with Pawel Sawicki April 2, 2022)

He went on to describe how this method was never going to be mathematically precise in terms of numbers, as he only posts the twelve daily names, but there is an attempt to present:

A Roma victim or a Czech victim. However, of course we try to keep this balance that the two major groups of people deported to Auschwitz, where Jews by far and Polish people. However, from the educational point, if it is important to show those various aspects of the story of Auschwitz to show that they were Roma people, that they were Soviet prisoners of war, that they were Jehovah witnesses, that they were a small group of homosexual men, that there were people of various nationalities. So, you will find some people of Spanish origin. (Zoom interview with Pawel Sawicki April 2, 2022)

And indeed, over the two years, I have seen photographs of Jewish Italians, Norwegians, and Moroccans, Jewish kids born in Tel-Aviv, and Polish Catholic nuns and priests. I want to explore how I have listened to these images, and what this Twitter interface has done to my understanding of these pasts. Welzer notes that "Acquisitions and applications of pasts always follow the needs and demands of the present" (Welzer 2010: 6), and a social media platform such as Twitter has been arguably a most effective communication application in the early twenty-first century. The museum has come out of its walls and into our phones and tablets, with the images, many of them signed by the subjects, center-stage, and anchored by those bare facts of date and place of birth, home, occupation, and fate: "she did not survive." There are three types of images that Sawicki posts: the triptych of the person once inside Auschwitz in their striped uniform and head shaved, some of the women with headscarves on, there are group photos and longer text explanations of resistances, and there is the "before" photo of "pre-war selves" (Hirsch and Spitzer 2009: 160); babies in cots, children on their first day at school, women in deckchairs, on the beach, "25 April 1891 | A Hungarian Jewish woman, Rachel Sajovics nee Geist, was born in Sajomagyaros. In June 1944 she was deported to #Auschwitz and murdered in a gas chamber," men in swimwear, "1 May 1903 | A French Jew, Léon Mozes, was born in Paris. He was deported to #Auschwitz from Drancy on 30 June 1944. He did not survive," and photo after photo of women posing for a studio camera, in their finery. They resonate, as Campt argues, because they talk to me (I don't assume any affect on others) of moments I can identify with because they look like people from my family albums. I am not Jewish, but I recognize my father as a child on a beach with a woman, my grandmother whom I never met, and the two of them share similar clothes and hairstyles to the Twitter feed. They come from the same times; images captured in the 1930s. They are somehow, a "shared memory" in this respect. In their writing on the ethics of memory, Margalit writes how this type of memory "in a modern society travels from person to person through institutions such as archives, and through mnemonic devices, such as monuments and the names of streets" (Margalit 2002: 54). Now Twitter is a mnemonic platform, and it is doing something affective to the memorializing process, as Kuhn articulates it: "Remembering is institutionalized through cultural means—in objects, material culture (monuments, books, and suchlike) as well as through practices and rituals of commemoration that may involve, but are not confined to, what participants actually remember from their own experiences" (Kuhn 2010: 1). This feed has reconfigured the way that I listen to this particular past. Growing up in the UK as a child in the 1970s, my recollection of encounters with newsreels on television about the Holocaust were full of images filmed by the British and Americans who arrived at Auschwitz at the liberation on January 27, 1945. I recall seeing large numbers of emaciated people with shaved heads wearing rags. This was how I received the information about this past, and how it was reiterated in my own children's school history books. I recall recoiling from the television when I saw those images, scared of all those people behind barbed wire. The footage was dehumanizing. The Auschwitz Museum's Twitter feed returns these people to dialogue, it reconnects them with the present, it vivifies them. But because of where they are, in the spelling out of where the photos were taken and the text that informs us of the inevitable trajectory from life to death, there is also, in parallel, a spectral quality to them which gives them the ability to "haunt" (Sontag 2003). Sontag pitted narrative against image, noting that narratives enable understanding, whereas photographs haunt. I dispute this binary, but it is being used to effect by Sawicki's strategy. On the @AuschwitzMuseum thread, while the text anchors and informs, the photographs sing out. They have a similar efficacy to what Tandeciarz noted in Argentina where the mothers of the *desaparecidos* silently paraded with photographs of their loved and lost ones:

> What is it about the irruption of women in the public sphere and, specifically, their use of photography to turn "their bodies into billboards [and] conduits of memory" (Taylor 2003: 170) that makes this method of resistance so poignant, so meaningful, and so transnational, capturing the imaginations of so many? (Tandeciarz 2006: 137)

For Tandeciarz, the question is related to resistance, that of the mothers refusing to let go of their children via their memory and asking others to remember them too. Here, in the @Auschwitz feed, the method might also be termed a form of resistance whereby individuals who have been absent are returned to collective memory via photographic images. Tandeciarz writes of how "The use of photographs to remember an absence re-creates, symbolizes, recuperates a presence that establishes links between life and death, the explicable and inexplicable. Photographs 'vivify'" (p.137). And this word gets to the core of the affective power of the feed. The photos live not just because of their own vitality but by "attaching a face and a name to the disappeared [murdered], these photos imbue them with a corporeality they have been denied; . . . they reinsert the missing in the spaces from which they have been torn" (p.138). I suggest they do something else as well. They exist, those twelve photos, above and below other images from any account that someone might be following. For me, these photos appear in a stream among academic, university, political and media accounts. They become part of the daily diet of images that I scroll through. They exist alongside these other images on the flat schizophrenic surface of my iPhone; postmodern and like early MTV (Kaplan 1987). I along with the other 1.3 million followers are fleeting witnesses to these people. This is an odd position because of the interplay of horror and banality, of the person and their fate and the platform where I am encountering it. Felman and Laub have written that the Holocaust was "An event without a witness" (1992: xvii). There is an "unspeakability" about it; words are not adequate, lying as it does, beyond representation. It could be that the museum's account on this social media platform goes some way to delivering these lives into representation and "decanoniz[ing] the silence" (Felman and Laub 1992: xix) by naming and vivifying through the resonant image.

In her research on music and memory, Istvandity refers to Halbwachs' view of how collective memory operates "to reconstruct an image of the past which is in accord in each epoch, with the predominant thoughts of society" ([1952]1992, 40 in 2019: 26). Writers on autobiography (Freeman 2010) and on memoir (Radstone 2007) have argued that these narrative modes are strategies for writing the self into history (Gardner 2020: 17). The @Auschwitz account might be read as a form of autobiography where people whose names were not broadcast beyond their immediate families are recuperated into a larger historical narrative on a platform that presents ideas and conversations in a very particular fashion.

Twitter is a "loud" digital social media platform, arguably the loudest, where its short form discourages nuance; instead, it is perfect for pithy or controversial takes. Algorithms tend to favor those with the loudest voices, who can make the most noise. What better place to insert these resonant, quiet, and dignified images whose whispers, cries, and laughter we share. Within Twitter, my own "echo chamber of cyberspace" (Burton 2005: 1) I bear witness via a connected listening. I can hear the stories of the murdered, of those whose fate was unknown, and of the few survivors amidst the cacophony of accounts tweeting out the banal, the useful, the promotional.

This online listening happens in a specifically twenty-first-century, bite-sized way, operating a kind of distracted engagement. I remember some stories more than others; I might screen grab an image and keep it to look at later, to really listen to the four lines and what they tell me. For Yoseph and his mother Dasza, I feel horror for their move to Europe. "20 April 1928 | A Jewish boy, Yoseph Wacholder, was born in Tel-Aviv, British Mandate for Palestine. He lived in Paris. He arrived at #Auschwitz on 7 August 1942 in a transport from Beaune-la-Rolande with his mother Dasza. They were probably murdered after the selection." This story has a texture to it, an emotional weft, it is "Matter . . . shimmer[s] with undetermined potential and the weight of received meaning" (Stewart 2007: 23), it carries so much in its forty words. I listen to this brief narrative and learn the names of people, many of whose deaths are unmarked and unknown, who were "murdered twice, both in body and in name" (Margalit 2002: 20). The @AuschwitzMuseum account is a naming device. It returns people as recognizable individuals embedded in family and in life events and so operates an affective familiarization whereby the listener is, again, brought close and able to listen to the stories of those who were silenced. What the Auschwitz Museum does is subvert the silence, give names to the boys, and girls, the mothers, sisters, aunts and uncles, the priests and soldiers, the resistance fighters and we see them, and not how they were documented. Or rather, we see them as their families documented them. We operate a familial gaze. This and the naming bring them closer. The @Auschwitz posts invite me to engage in an affective, temporary listening attachment to the "resonant" image of the "historically dismissed" (Campt 2017: 11). I hover between remembering and forgetting, which is a "shorthand for a constellation that might include denying, ignoring, refusing, disavowing and being indifferent to" (Moreno 2016: 158), but when I stop and listen, it is to the image of the person who was born on the day, which leaves a deep "impression" on me (Campt 2017: 72). This engagement is afforded through a connected listening that invites in memories of my own past, my families. This connection is further compounded through a "grammatical" (Campt 2017: 59) familiarity that draws on personal memories of family photo albums as well as a collective knowledge of the history of the camps and the Holocaust. Like the Veterans' Voices, it is the ordinary that takes precedence, "the production of the ordinary. The ordinary is here fantastic" (Ahmed 2004: 118). Stewart's poetic comments on the ordinary go quite a way to conveying the nature of the online interactions that this Qualitative Online Listening has afforded me, at least. She writes: "The ordinary throws itself together out of forms. Flows, powers, pleasures, encounters, distractions, drudgery, denials, practical solutions, shape-shifting forms of violence, daydreams and opportunities lost and found. Or it falters, fails. But either way, we feel its pull" (Stewart 2007: 29). Sawicki tells me about this, about scale and the insertion of the "small." He talks about "those small, tiny memories because I believe that The Auschwitz memorial is not one memory. There are many memories about Auschwitz and our role is to acknowledge all of them and preserve all of them, even if it's not perfect in some sense." He acknowledges that they cannot include everyone who went through the camp and was born on a specific day because of the specific logistics that confronts him:

> I would have to then have 100 faces a day or 500 faces a day to be able to. However, it's impossible, it's, it's too much. It's challenging and of course choosing . . . sometimes it's the question of availability of the sources . . . we can reach the data which we can check whether because sometimes when you get to some archives you have a face, you have a name, but you don't have a full birthday or you don't have any information about the date of the transport. (Zoom interview with Pawel Sawicki April 2, 2022)

Conclusion

Listening across Age(s) involves listening with an awareness of the modalities and context that age affords. It is also about listening to the past as it appears in the present, in new formats and in affective narratives. In this chapter, the Veterans' Voices project illustrated the types of listening that happen with older people, and how performative historiography offers them and their listeners a map of the past that is generally agreed upon through use. Narratives became sedimented through reiteration. The past was scripted in a certain way. Listening across Age(s) can also be an upsetting experience when that past appears in newly formatted guises. This can be in format, both Digital Stories and Twitter are both short-form media. Digital Stories are (historically/traditionally/usually) two to five minute long. Twitter only allows for 280 characters a tweet. The limitation of the platforms dictates the scale of delivery and where they might appear. These small media are mobile and can pop up and insert themselves into the lived experience of the everyday. This is especially true of Twitter, where the Auschwitz Museum's short stories of people murdered in the Holocaust might be listened to on the train or the tube, or when you are scrolling for other information. They insert themselves into the "affective economy" of social media, whose responses are dictated by the like or retweet button. Listening to these pasts as they are reformulated is also upsetting in terms of content; repressed content is released, the ordinariness of the murdered acts as counterweight to the mass-ness of history. I have listened to details, "3 mates" being bombed in a jeep in Aden, now Yemen, the good wages earned while doing National Service, the ordinary life events of individuals whose names I can remember, Yosef, Genevieve, and Suzanne. These listening interfaces have produced different stories about the pasts of Aden and Auschwitz. These stories now add to the echoes that sound out across our shared corridors of ages past.

4

Listening and Belonging

Permission to be heard.
Can I speak?
Yes.
But who is Listening?
Can I speak?
No
We don't want to listen to you.

The Old English word for "belonging" is "be-longen," which means suit, or pertain to. "Longen" comes from the Proto-Germanic "langian" (sometimes "longian"), which in turn means to yearn for, long, or grieve. So, wrapped up in the contemporary version of the word are remnants from the past that indicate how belonging has been about yearning and loss. It is something we might both strive for and mourn, it is always before and behind us, we are forever on its edge(s).

Here I explore the relationship between listening and belonging, and how it is key to understanding contemporary British citizenship. Using examples from historical and contemporary BBC Radio, funded digital storytelling research that I have conducted with people on the edge of work, and observations on current court cases and UK government policy where listening has been denied, I listen at some of those edges. This is where the skirmishes are, because edges are like borders, which I deal with in the following chapter. Here I focus on the notion of the "cultural" citizen (Avelar and Dunn 2011: 3) to look at how cultures of belonging are produced within affective regimes that prioritize the familial, which I argue are found in the BBC Radio broadcast from the early part of the twentieth century (*Listen with Mother,* 1950–82) and a contemporary listening program (*The Listening Project,* 2012–22). The BBC is Britain's Public Service Broadcaster, established in 1927 under the leadership of Sir John Reith as director general (1927–38) and whose mission was to educate, inform, and entertain. I have used this source rather than commercial channels given its historical and continuing position "as" the national broadcaster.

I find it useful to apply Berlant's idea of "ambient citizenship," which she clarifies as being "a mode of belonging, really, that circulates through and around the political in formal and informal ways, with an affective, emotional, economic, and juridical force that is at once clarifying and diffuse" (2011: 230). The BBC uses this affective and emotional mode to generate ideologies of belonging with respect to place; a woman's place in the home (in *Listen with Mother*) and the tensions over Britain as narrated by people who have moved here (*The Listening Project*). In particular, it focuses on the sixty-seven conversations collected in the British Library's archive collection of the project on the subject of migration, in order to flesh out the idea of "ambient" citizenship, in terms not only of "feeling" and "environment" (ambience) but of movement ("ambient" Latin meaning of "going around"). "Movement" is considered more in the following chapter in relation to music, storytelling, and belonging; here I am interested in the shifting borders of citizen and non-citizen (Braidotti 2013), how noise is being used to police that boundary and how listening and not listening function to configure that particular border. I think about "non-citizens" here as people who are too noisy to fit into that categorization; people whose "noise" comes from other places and positions, be that the noise of the empire returning, or the noise of dissent to power. So, while being in broad agreement with Jones' definition of the term, which they clarify as "from its early origins through to the present day, citizenship has always been a two-sided concept, with the state guaranteeing rights for some people while excluding many others from the right to have rights" (Jones 2017: 79; Bradley and De Noronha 2022), I seek to focus on the listening spaces where this particular type of belonging has been and is being enacted.

There are also policies and strategies of "not listening" to specific individuals or groups of citizens, which render them beyond the bounds of the heard, because they somehow breach the "ordinary." Throughout the chapter, a "fantasy of ordinariness" (Berlant 2008: 11) appears as rhetorical recognition, with an interpellative function, that also works to construct borders of belonging, that invites, expels, and discriminates. Berlant's work focuses on America and its conservative politics, and their focus is on sexuality, bodies, and the "intimate public sphere" in relation to citizenship. They write that "citizenship, in its formal and informal senses of social belonging, is also an affective state where attachments that matter take shape" (Berlant 2011: 163). My emphasis is first on these attachments that are formulated via the constructed "familial" intimacies afforded by moments of connected listening across state radio and in the digital storytelling elements of a funded project I worked on, which, I argue, is about the production of a "good" (i.e., productive) citizen. Second, I look at processes of defamiliarization, which deny listening through the application of "noise" and discuss recent examples from the UK where other and oppositional bodies are expelled from the body politic because they are too noisy (Thompson 2017).

Listen with Mother

Modeling their radio program on a bed night reading routine, *Listen with Mother* was first broadcast in Britain on the BBC Radio's Light Programme on January 16, 1950, continuing the same format for thirty-two years, with a move to BBC Radio 4 on September 30, 1967. The description of the program from the BBC's own website[1] reads as follows:

> *Listen With Mother* was first heard on 16 January 1950. It offered a mix of nursery rhymes, stories and music for the under-fives and their mothers but over the years developed a following across the generations. Each episode ended with the Berceuse from Faure's *Dolly Suite*, played on the piano by Eileen Brown and Roger Fiske.
>
> *Listen with Mother* was broadcast at 1.45pm when children would be ready to concentrate after their lunch, and mothers would have time to sit with them. The presenters—including Daphne Oxenford, Julia Lang and Dorothy Smith—adopted a new intimate tone, talking as though to each child alone. The centre of the programme was the story, preceded by the calming phrase "Are you sitting comfortably? Then I'll begin." The question, originally an ad lib by Julia Lang, became so well known that it ended up in the *Oxford Dictionary of Quotations*.

The description of the program conjures up a picture of a 1950s' British household, where the (married) woman has just cleared up a lunch she has shared with her child/ren, with whom she now settles down in a cozy armchair or sofa to listen to the broadcast. Maybe a nap will be had after the show. Father is not there, nor are any other extended family members. Mother is most possibly married to a man. The direct address mode ("Are you sitting comfortably"?)

---

[1] https://www.bbc.com/historyofthebbc/anniversaries/january/listen-with-mother/

was the voice of a surrogate (radio) mother transmitting to the domestic sphere. And as the BBC said, the three female presenters' "new intimate tone" was a shift from the informing, educating, and entertaining voices of BBC news and light entertainment. This was the voice of the mother, enabling the "real" mother to take a step back and hand over control of her offspring to someone else. What a relief...

Timothy Taylor writes about intimacy in relation to early American radio broadcasts and his point about audience works well to understand not only *what* Listen with Mother offered, but *how*

> the intimacy that was thought to be intrinsic to radio because of its placement in the home resulted in an unprecedented intrusion into peoples' private spaces and lives. The focal point of this intimacy, however, was not all listeners, but women. (Taylor 2005: 261)

Although scheduled for early afternoon, just after lunchtime, it was modeled on the Victorian idea of the bedtime story (1873 in Louise Moulton's novel, *BedTime Stories*). This ritual was an enculturing process that was performed by (some) parents (Heath 2000). It had a pedagogic and literacy goal; to help young children read, and by reading, to be more able to proceed through an education system and develop into a fully functioning adult citizen. Part of a model nuclear family daily ritual, the bedtime story would be read after the bath/washing of the teeth, to settle the child, and to make reading part of the map of the everyday. Only high days and holidays, early departures, sleepovers, and parties might be distinguished as exceptional because of the lack of a bedtime story. It and its role in the "family," reproduced in *Listen with Mother*, meant that it was part of an accepted routine of normal family life, where normality was a married woman at home looking after children under five years. It was where "listening as a modality of domestication" was being fomented (Kheshti 2015: 40, 3).

As of June 2022, there were 1,185 episodes of the program from 1950 to 1982 available online.[2] They are not in chronological order. For the first decade, the archive includes descriptions of the program content, drawn from *The Radio Times*. This is from August 25, 1952.

> It is not only our "under-fives" who sometimes realise a loss when they go to school and have to give up their programme; it may happen to mothers, too. "Much to my surprise," writes one, "I found a big lump in my throat when I

---

[2] .https://genome.ch.bbc.co.uk/search/160/20?q=Listen+with+Mother#top.

happened to turn on the programme one day. The sudden realisation that these happy times of listening together were over came as a great shock." But it is happy feature of our young audience that, as some of its members move away, others are drawn in. There can hardly be a day when the attention of some very small child is not caught, for the first time, by some part of the programme. Thus, mother finds herself again "rushing to tune in at 1.45."

The stories told across the thirty-two years were what might be expected of stories for the under 5s, humorous and instructional tales of animals and adventures. But some stories did overt ideological work. On July 31, 1958, "today's story was about 'The Sunday School Outing' by E. R. Hutt, told by Dorothy Smith." This was arguably a conservative post-war decade and

> Sunday Observance was strict—even the swings in the park were tied up on Saturday nights so that no one could use them on Sunday. Shops were shut and sporting events were banned. Most churches were relatively full, and three quarters of the nation's children went to Sunday school. All of the nation's children had a daily period of worship and Bible based teaching every week.[3]

Britain is, at this point, a homogenous Christian country. The absence of stories about Passover or Yom Kippur (from the 1950s to 1982), to Eid (from the 1970s to 1982) meant that certain British citizens' heritages were not storied. Other ideological work is done in relation to expectations around gender and domesticity, where children might help their mothers in the house, but the outside activities are led by their fathers. In 1980, the anniversary of the British Queen Mother was used as a route into broadcasting stories about "grannies," one of a number of vernaculars for "grandmother." The familial ties between their own families and the British Royal family are established as accepted. The children are listening not only with their mothers, but together in a listening community of mothers and their young children.

What better way to underline gender expectations and indicate the timetable of the domestic day's labor by carving out this listening space; a respite from mothers who could relinquish control to the radio mothers. And just as "radio became a prime site for the establishment of national identity through national culture" (Hartley 2000: 156), so the exhortation to "listen with" brings with it an intimate listening moment that also reiterates myths of motherhood as it is indelibly related to domesticity. Along with the stories about animals, a number of broadcasts were

---

[3] https://christianconcern.com/wp-content/uploads/2018/10/CC-Resource-Events-2020-Gospel-Issues-Clifford-Monica-Hill-Why-Is-The-UK-Church-Declining-Notes.pdf

themed around domestic chores. A selection taken from the site (on June 6, 2022) shows how certain domestic labor was included in the stories: washing clothes, washing houses, washing dishes—all domestic chores recast in story mode. Fathers appear outside this domestic space ("days out," "walks with Dad").

> "Helping Mother" December 14, 1962
> "Grandpa's Surprise" December 8, 1965
> "Nicolas Helps with the Washing Up" 24 January 1966
> "A Walk with Dad" November 15, 1971
> "Out with the Dad's February" 3, 1972
> "Going Fishing" April 7, 1975
> "Spring Cleaning Day" March 19, 1975
> "The Different Wash-Day" September 29, 1975
> "Bath Time for Ben" January 5, 1976
> "Granny week. A special week of granny stories to celebrate The Queen Mother's 80 birthday" August 4, 1980

In her work on the intimacies that weave their way through personal and political bodies, Lauren Berlant argues that

> there is no room to make a distinction among political, economic, and affective forms of existence, because the institutions of intimacy that constitute the everyday environments of the social are only viscerally distinct but actually, as we know, intricately and dynamically related to all sorts of institutional, economic, historical, and symbolic dynamics. (Berlant 2011: 168)

Listen with Mother is intimate in tone, format, and medium. And as Berlant argues, it is "dynamically related" to other broader dynamics, that is, those of a mid-century model of maternal identity that foregrounds domesticity and heteronormativity. It is placed, in mode and time, as a listening practice whereby ways of being British and female are relayed through story. The father figure of the story is not part of the "we" (*Listen with . . .*) but invoked as someone absent. No fathers are included in any of the Radio Times' descriptions of programs that appear in the British Library archive from the 1950s are like radio shout-outs in contemporary broadcasting ("'at the beginning of this series' wrote a mother recently," July 24, 1950; "Our correspondence since these programmes began." July 3, 1950). This is an early example, broadcast on February 2, 1953:

> "I teach music in a rural area," writes a correspondent. One little girl of five has only begun school this last half term. When we sang several nursery rhymes as

revision, I noticed she was word-perfect. I asked if her mother sang to her at home. She said "No, I listen every day to Listen with Muvver". Perhaps this little girl, and many like her, may hear Dorothy Smith today and on Thursday and Friday, telling them about "Taddy's Tail" by Jane Shaw, and "Blackie the Coal Engine," by Mary Cockett.

These snippets in *The Radio Times*, a radio listings magazine first published in 1923, sutured the domestic lunchtime units of mothers and children to others across the country, linking them further to schoolteachers. The program managed to establish "listening with" as a process of "connection to." This connection was one which foregrounded the essentially ordinary but only way of being as a mother and child in 1950s' Britain. It manufactured a shared sense of belonging via its scheduling; as Garde-Hansen and Gorton note in their work on television and memory (2019) when they acknowledge Needham's (2009) ideas on time, who uses Judith Halberstam (2005: 5) arguing that they "characterize normative time by drawing attention to the different categories through which it comes to be organized and given meaning. These include the interconnectedness of family time, reproduction time and inheritance time" (2009: 151; our emphasis 2019: 3). These moments encapsulate how radio enabled a type of recognition of belonging via listening which emphasized a shared normality, "of things deemed normal, an order of what was felt to be a general everyday intimacy" (Berlant 1997: 2) in both form and content.

One from March 3, 1953, notes how affective the listening process can be in terms of moving stories into the listeners' lives:

> "Best of all," wrote the mother of a two-and-a-half-year-old boy, he liked "Susie is a Monkey." We had to replace Susie by all the little boys and girls we know. "And this one" he said finally, pointing to himself.
>
> We have many such pleasant examples of how our stories and rhymes live for the children far beyond the moments of hearing them, of how they enter into their lives and go on nourishing them in often unforeseeable ways. and it is one of our great satisfactions that out of the wireless set, which presents the same material to so many children, can come so much variety and individuality of response, and that each listener so easily adapts it to his own needs.

The program is therefore a good example of the replication of maternal/child intimacy as a moment shared via a modern technological medium of the mid-twentieth century in Britain. The belonging afforded in those listening moments happens in a dyadic relationship; mother-child unit to radio and the mother-child unit to itself. These listening bubbles, broadcast into British houses at 1.45 for thirty-two years, inculcated the importance of the (Christian) mother-child relationship within that domestic space. This is where they belonged. Over fifty years later, the BBC is broadcasting conversations that are determined by "listening" to the intimate, the familiar, and the ordinary.

The Listening Project

The Listening Project broadcast on the BBC from 2012 to 2022, finishing just as this book was in copyedit. It was less a broadcast radio show, rather, it offered a series of "conversations" that can still be downloaded on a variety of contemporary topics. Local BBC Radio producers worked to generate the show and the edited conversations appeared (and remain) on the BBC site, introduced by radio journalist Fi Glover. The unedited ones were then archived in the British Library "for future generations to hear. So far over 600 conversations have joined the British Library's extraordinary collection of over 6.5 million sounds, where they are available to everyone for research, enjoyment, and inspiration."[4]

On its website, it foregrounds the idea of the intimate: "The Listening Project is the BBC initiative in which people across the UK record a conversation with someone close to them on a subject they've never discussed intimately before."[5] This statement clarifies it further: Radio 4's Controller Gwyneth Williams said: "We want to bring a new kind of conversation onto Radio 4, made possible by the unique nature of radio, its intimacy and entanglement in the lives of audiences."[6]

This intimacy is further compounded by the mobile Listening Booth, modeled on an Airstream caravan, that traveled around the country to people, rather than having people come to a local radio station and be recorded. The speakers were introduced with relevant biographical details, age, and so on and then left to converse. The BBC described it as:

> an audio archive of conversations recorded by the BBC. People are invited to share an intimate conversation with a close friend or relative, to be recorded and broadcast (in edited form) by the BBC and curated and archived in full by the British Library. These one-to-one conversations, lasting up to an hour and taking a topic of the speakers' choice, collectively form a picture of our

---

[4] https://www.bbc.co.uk/mediacentre/latestnews/2015/listening-project-booth
[5] https://www.bbc.co.uk/mediacentre/latestnews/2015/listening-project-booth (accessed June 13, 2022)
[6] https://www.bbc.com/mediacentre/latestnews/2012/listening-project (August 25, 2022)

lives and relationships today. Oral history recordings provide valuable first-hand testimony of the past. The views and opinions expressed in oral history interviews are those of the interviewees, who describe events from their own perspective. The interviews are historical documents and their language, tone and content might in some cases reflect attitudes that could cause offence in today's society.[7]

On the occasion of its tenth anniversary, on June 10, 2022, its BBC Gloucestershire producer, Faye Hatcher, tweeted: "**Work request** For The Listening Project's 10th anniversary @BBCRadio4 I'm looking for 2 chatty people born in 2012, to have a natter about what the world might look like in years to come. Your conversation will be archived @britishlibrarytoo (15.00 GMT, 10.6.22)."

This establishes the program's archival goal and even advertises it as an incentive to participants. Their voices can contribute to, what Fi Glover, its host, describes as "this nation's bank of experience and memory."[8] These contributions are couched in the terminology of capture, with its intimations of control and seizure, they are described as "Capturing the nation in conversation"[9] like Victorian-era butterflies; a "unique portrait of the UK."[10] They provide an audio collection and I want to think about that in relation to listening and belonging. In 2016 I wrote a paper with my then colleague, Gerry Moorey, about the trend to "rediscover" artists, and we used Appadurai's 1986 piece about collecting and colonialism. We argued that it was "useful to look back to Appadurai's (1986: 46) work on commodities and cultural perspectives, where they reiterate how, within a long tradition of (colonial) collecting, collectors are driven by the 'obsession with the original' within what they term a 'cultural regime of authentication'" (Gardner and Moorey 2016: 173). The BBC's *Listening Project* is aiming to do that, to bank real people's voices and collect them as part of an oral history endeavor. Like making a memory box and burying it, these voices talking about current crises and politics, and about themselves, are the national broadcasters' attempt to weave the ordinary "citizens" into an aural map of early-twenty-first-century Britain.

Clicking on the program's website, I listened to the main reporter, Fi Glover, introduce an episode of four conversations about the cost-of-living crisis, travel chaos at the UK's airports, Northern Ireland's future, and bushcraft skills. She starts by saying that "It's probably only by listening to other people's stories that

[7] https://sounds.bl.uk/Oral-history/The-Listening-Project#
[8] (https://www.bbc.co.uk/mediacentre/latestnews/2015/listening-project-booth).
[9] https://www.bbc.co.uk/programmes/m00187pc
[10] https://www.bbc.co.uk/mediacentre/latestnews/2015/listening-project-booth

we can ever reshape our own." She is alluding to the dynamic nature of listening (Lloyd 2009; Kassabian 2013; LaBelle 2018) where "mirroring" (Illouz 2007: 20) enables identification and engagement.

On the British Library site, the conversations are grouped into different topic areas; "Community," "Education," "Family," "Friendship," Health," "History," "Loss," "Love," "Migration," "Politics," "Religion," "Sexuality," "Sport," "War," and "Work." Because the next chapter is about a project whose participants were migrants, I want to concentrate on the sixty-seven conversations collected in the British Library on the topic of migration to focus on the listening that is happening in these instances.

- Conversation between asylum seeker and support volunteer, Anwar and Dorothy, about their involvement with the Support for Wigan Arrivals Project. 00:40:08
- Conversation between brother and sister, Hector and Delores, about moving to the UK from Jamaica in the 1960s. 00:28:22
- Conversation between brother and sister, Mohamed and Najah, about coming to the UK from the Dadaab Refugee camps after fleeing persecution in Somalia. 00:15:31
- Conversation between brothers, Dominic and Jamie, about leaving Liverpool to come and live in Derry, Northern Ireland. 00:59:02
- Conversation between brothers, Tawona and Ernest, about moving from Zimbabwe to Glasgow. 00:50:55
- Conversation between colleagues, Adam and David, about being migrants in London and what they love about the city. 00:39:06
- Conversation between colleagues, Cath and Mick, about working with refugees and asylum seekers at the Support for Wigan Arrivals Project. 00:41:34
- Conversation between employer and employee, MD Atikul and Hasin, about moving to the UK from Bangladesh, their supportive local community and experiencing racism. 00:36:02
- Conversation between father and daughter, Carson and Ashley, about Ashley's experience of growing up in Cambodia. 01:02:02
- Conversation between father and daughter, Dixie and Anya, about Anya's recent move to Switzerland to start a new job. 00:45:36
- Conversation between father and daughter, Maris and Nikola, about why Maris left Riga, where Nikola still lives, to live and work in Guernsey. 00:47:19

- Conversation between father and daughter, Peter and Catherine, about their family connections with America and Northern Ireland and their memories of the Troubles. 01:17:58
- Conversation between father and daughter, Russell and Joanne, about why Russell chose to leave his London life behind and move to Lincolnshire to raise his family there. 00:43:55
- Conversation between father and daughter, Ulrich and Anna, about Ulrich's German background. 00:25:22
- Conversation between father and son, Jalal and Salwan, about Jalal's experience of growing up in Baghdad and then moving to Belfast. 00:45:44
- Conversation between father and son, Marwan and Joudi, about coming to the UK as refugees from Syria. 00:24:20
- Conversation between father and son, Prakash and Sunil, about how Prakash was forced to leave Uganda and move to the UK in 1972. 00:44:23
- Conversation between friends and colleagues, Hector and Aleks, about architecture, London and their idea of home. 01:15:43
- Conversation between friends and colleagues, Madara and Agnese, about moving from Latvia to Guernsey for work. 00:55:28
- Conversation between friends, Aleona and Emilia, about being newcomers to Shetland and how they have fallen in love with its landscape and its people. 00:56:21
- Conversation between friends, Andy and Andrew, about how Brexit might affect them as English expats living in France. 00:42:45
- Conversation between friends, Ashelder and Kuncil, about volunteering together at the Support for Wigan Arrivals Project. 00:37:11
- Conversation between friends, Auliya and Nawa, about Auliya's experiences of growing up in war-torn Afghanistan and seeking asylum in the UK. 00:42:12
- Conversation between friends, Beth and Misa, about how Beth supported Misa when she first arrived in Shetland as a student from the Czech Republic. 00:48:59
- Conversation between friends, Elsie and Netta, about how they miss Jamaica despite having lived in the UK for many years. 00:51:07
- Conversation between friends, Frank and Hubertus, about both being German and living in Guernsey. 01:26:51
- Conversation between friends, Gulay and Comfy, about their experiences of coming to the UK to study and adjusting to life in a foreign country. 00:36:31

- Conversation between friends, Heather and Kathy, about their friendship and moving to Shetland from elsewhere. 00:43:03
- Conversation between friends, John and Jill, about the work they have done with refugees around the objects they have been able to bring with them to the UK. 00:32:10
- Conversation between friends, Kasey and Georgia, about moving to Belfast from California and London. 00:47:59
- Conversation between friends, Linford and Whit, about coming to Britain as Jamaican immigrants. 00:53:29
- Conversation between friends, Muleya and Mubina, about coming to Scotland as refugees. 00:50:46
- Conversation between friends, Patricia and Meriem, about seeking asylum in Scotland. 00:54:11
- Conversation between friends, Ruth Schwiening and Ruth David, about coming to the UK as Jewish refugees on the Kindertransport. 00:52:08
- Conversation between friends, Tarsem and Gurvinder, about Tarsem's experience of moving from India to England, arranged marriages and being British. 01:04:01
- Conversation between grandfather and granddaughter, Fred and Jhanae, about how different school was when he was growing up in Jamaica. 00:52:52
- Conversation between grandfather and granddaughter, Hargun and Priya, about Hargun's experience of migrating to Britain from India in the 1950s. 00:04:44
- Conversation between grandfather and grandson, Brian and Charlie, about Brian's experience of growing up in Brazil in the 1930s. 00:47:38
- Conversation between grandmother and grandaughter, Ndaizivei and Sekai, about Ndaizivei's experience of escaping from Rhodesia as a political exile. 01:27:37
- Conversation between grandmother and granddaughter, Olivia and Olivia, about growing up in Jamaica and coming to live in Britain in the 1950s. 00:54:57
- Conversation between husband and wife, Bernard and Kath, about their plans to emigrate from Llandudno to Sydney. 00:30:01
- Conversation between husband and wife, Bradley and Pippa, about moving to England from South Africa. 00:42:46
- Conversation between husband and wife, Javaid and Shabana, about their memories of Kenya. 00:38:05

Listening and Belonging

- Conversation between husband and wife, Kenneth and Patricia, about their very different childhoods and spending most of their married lives living an ex-pat lifestyle in India, Pakistan and Kenya. 00:54:22
- Conversation between husband and wife, Mike and Pip, about emigrating to Spain. 01:14:31
- Conversation between husband and wife, Stephen and Laura, about their grandson who was born in Australia because their son has emigrated there with his wife. 00:54:05
- Conversation between husband and wife, Tico and Sharon, about moving to the UK from Canada and what it means for them to be British. 00:23:38
- Conversation between mother and daughter, Ade and Kitty, about Ade's Nigerian roots and heritage. 00:43:49
- Conversation between mother and daughter, Alice and Michelle, about Alice fleeing to England from Zimbabwe. 00:52:25
- Conversation between mother and daughter, Celia and Mary Jane, about Celia's childhood in Rhodesia. 00:47:29
- Conversation between mother and daughter, Cynthia and Melanie, about Cynthia's memories of growing up in India during the final days of the British Raj. 00:13:25
- Conversation between mother and daughter, Ivaline and Joy, about Ivaline's experience of moving to London from Jamaica and the feeling of belonging. 00:26:16
- Conversation between mother and daughter, Nancy and Pearly, about the possibility of Pearly needing to migrate again. 00:49:16
- Conversation between mother and daughter, Polly and Victoria, about Polly's move from Jamaica to Reading in the 1960s. 00:46:00
- Conversation between mother and daughters, Aimee, Anne and Anielle, about their new life in Scotland after gaining refugee status and leaving the Ivory Coast. 00:26:01
- Conversation between mother and daughters, Alison, Keira and Raegen, about Alison's imminent move to work with the church in Australia. 00:43:45
- Conversation between mother and son, Brigid and Patrick, about Patrick's experience of working in Australia. 00:41:02
- Conversation between mother and son, Gemma and Callum, about moving to Northern Ireland after the recession in Galway and the potential impact of Brexit. 00:49:31

- Conversation between mother and son, Halina and Thomas, about Halina's Polish background and being sent to England for a better life. 00:54:51
- Conversation between mother and son, Katy and Robin, about their Russian background and how it has influenced their lives. 00:45:36
- Conversation between mother and son, Lili and Danny, about how Lili escaped from Berlin during the Second World War and her relationship with Danny's father. 00:48:01
- Conversation between mother and son, Sonia and Juan, about their memories of Chile and how they escaped and were granted political asylum in the UK. 00:46:19
- Conversation between mother-in-law and daughter-in-law, Michele and Fanzi, about how Fanzi moved alone to England from China when she was 16. 00:46:55
- Conversation between partners, David and John, about their move from the city to the country. 00:26:51
- Conversation between partners, Mariusz and Aleksandra, about moving from Poland to Belfast. 00:39:31
- Conversation between sisters, Sara and Jane, about living in different continents and the need to connect face-to-face with each other. 00:40:06
- Conversation between sisters, Sheila and Christine, about growing up in Rhodesia in the 1950s and 1960s. 00:33:40.[11]

Most of the conversations collected here are about movement. There are forty-one conversations on the topic of the migratory move itself, that is, the story about coming from one place to another. Ten conversations focus more on the past, in terms of the experiences of growing up somewhere else. Two conversations are about the experience of being migrants, and three discuss working with refugees. The remaining twelve dwell on memories of childhoods (three), discussions over roots and heritage (two) while there are ones on family connections (two) and people talking about what they love about where they live now, how they feel supported, how they miss their "homeland" and how Brexit has affected their status abroad. There is then a push toward the actual activity of migration that is the substance of these conversations and many of them revolve around the "loss" or "yearning" that the Old English "belonging" invokes. They are conversations about reminiscence, difficulty, and exile.

[11] https://sounds.bl.uk/Oral-history/The-Listening-Project# (accessed June 6, 2022)

It is worth noting how the archive is presented. Each conversation is "between" two people, either family members, friends, or work colleagues. A large number are intergenerational "betweens" (mothers and daughters, grandparents, and children). Robyn Fivush, writing up empirical research with families from a memory studies and psychology perspective, highlighted the interconnection of narrative with family experience. They write that "in adolescence, we see the beginning of a life narrative that links events across time and places the self in relation to others, embedded in an unfolding human drama of interconnected stories" (Fivush 2008: 54). This relationality, they go on to argue, "provides a framework for understanding oneself as a member of a family that extends before one's birth and provides the stage on which one's individual life will be played out" (Fivush 2008: 55). The *Listening Project*'s conversations underpin that sense of self within family, both again, by format (conversations with) and topic (here, migration and memories of people and places left behind). The emphasis on the familial and intimate produces this series of half-hour edited conversations that, though recorded in what might be considered some form of public sphere, where exchange of views is aired and disseminated.

However, these voices are not political, they go nowhere in that sense, they do not impact on policy decisions, they do not build up and create any form of crescendo around the topic of migration. If they complain, their complaints are voices within dramatic arcs and documentary paradigms. They are intimate, they may be affective, they are logged and listened to, but they do not reverberate beyond the BBC's site, the participants themselves, and the British Library archive. An optimistic reading of the program would see it as encouraging "new forms of *inter*twinement," a concept that comes from Sevenhuijsen's work on the intersections of care, feminism, and democracy. They use Bovens' (1995) work on the "relocation of politics" (in Dutch society, Bovens 1995) to suggest that this affords a realignment whereby democracy might function in pluralities, in many different spaces. Sevenhuijsen writes how within the "public sphere [where] people will exchange narratives of what counts in their lives and become acquainted with the stories of others. In this way they will arrive at systems of shared meanings that will make sustainable forms of co-existence possible" (Sevenhuijsen 2003: 180). This model relies on the idea of identification and exchange, another way of considering how connected listening might be an aperture into new forms of democratic alignments and affiliations. *The Listening Project* is similar to this exchange model, and Faye Hatcher, one of its producers at Radio Gloucestershire noted the following when I had asked her what impact being involved in *The Listening Project* had on people:

> Tricky to answer, as I've never been a contributor, but the feedback I receive as a producer is almost, always extremely positive. Contributors tell me that being "allowed" to speak freely without interruptions on a subject their passionate about, is refreshing and liberating. Some people find the process cathartic, if talking about a personal trauma. I think there's a lot of personal satisfaction of telling a story how it really is. I think a lot of people feel proud their conversation is archived in the British Library and also broadcast on BBC Radio 4. Some people have said it's life-changing and they've taken a different direction in life, after contributing to the project.

And asking here about the conversations being stored in the British Library she responded:

> It's really important and at the very core of what we're about. We're not interested in talking to high-profile, well-known people who already have a platform, but "ordinary" people who wouldn't normally have a voice on mainstream media. It's important that conversations are authentic as possible. We encourage contributors to use their own language and to speak how they would normally talk in conversation, so we often hear a range of accents, slang and swear words. In our recordings we often hear references to technology, brands and trends; the stuff that is usually air brushed out of interviews and documentaries. It's a social history archive. As producers, we're trying to capture a snapshot of modern Britain, so future generations can listen back to what life was really like in 2022, though the project has been going since 2012. It's the biggest sound recording of its kind in this country. We know University academics, writers, therapists use the collection for research and case studies.

The replies circle around the importance of the "ordinary" and how that builds into an archive for use by varied professions, academics, therapists, and so on. It stitches the voices together to make a kind of quilt of diverse voices all grouped under "Britishness" and ordinariness. It functions like the @AuschwitzMuseum's Twitter account within an affective broadcast economy, prizing the intimacy of emotion and connection generated by these listening bubbles it carves out. Listening at this micro level (Hrycak and Rawakowicz 2009), within an institutionalized broadcast and library culture, is about flagging up the type of belongings that can be mapped out "as" British. They clarify and diffuse (Berlant 2011) who is "in." There is no invitation to the extraordinary, to the weird, or to the eccentric. This rhetoric of ordinariness underpins and dictates inclusion and invites engagement. What follows next is a reflection on a project where the prize is the ordinariness of work, of being employed.

Employed Listening

GEM (Going the Extra Mile) is a UK National Lottery Community Fund and European Social Fund-supported project[12] whose aim is to enable people who are far from employment, to move, with support, into education, training opportunities, and work. It covers the county of Gloucestershire, in the southwest of the UK and multiple partners are networked into the project, from upcycling charities, the Gloucester Deaf Association to GARAS, Gloucestershire Asylum, and Refugee Association. I was part of a team whose job was to monitor and evaluate the project (see Courtney et al. 2020), and I used digital storytelling methods to do so. I was working with academics whose research was in socio-economic impact and policy, who were based in a research center, whose focus is on place and (largely) rural and agricultural matters called the Countryside and Community Research Institute (CCRI).

After a number of meetings, they agreed that digital storytelling seemed to be a creative way through which they might gauge the impact this long-running project had had on people who had taken part in it, and from there, they asked "Navigator Developers," a network of professionals connected to its component charities and businesses across the county, to see if any of their GEM participants had a story to tell about how their lives had been impacted on by being on the scheme. In an evaluation report (2018), there had been some frustration about how to capture impact, and digital storytelling was adopted by the team at the CCRI and the University of Gloucestershire to open up a different avenue for impact capture. My job was to enable them to tell their stories of how being on the GEM project had helped them, and to co-produce digital stories. The methodology I used actioned the principles of care and listening that I outlined in Chapter 2 and were very much produced alongside the three individuals with the help of their support workers.

In July 2019, I outlined the digital storytelling process and what was needed in terms of time and materials, to a group of GEM managers and Navigator Developers, and they then had time to get back to the people they were

---

[12] https://glosgem.org/about.php#page-top

supporting to see if there was any uptake. By the autumn, three people were interested. One had a history of substance addiction, another had mental health challenges, and the third had been exposed to crime at an early age. All of them agreed to see me, with their support worker, to learn more about the process. I invited two participants who were being supported by the same navigator to have coffee and cake with me at a charity run café one afternoon in September, so we could get to know each other. I showed them some digital stories from an Erasmus+ funded project called "My Story"[13] that I had been part of and asked them to think about sending me some photographs so we could then work on scripting their story. The same process happened for the third participant. We recorded the audio in October and November, with me traveling to a shared space in Stroud to work with participants 1 and 2, and at home (because our chosen café was shut) with the third.

The first participant had sent me photographs that they had taken of a dry-stone walling course they had been on, and of some stained-glass paintings, which he had painted. GEM had organized for him to go on the stonewalling course, and he recounted how he had taken leadership of the team on the day and had gained top marks; he had a photograph of his certificate and remarks how being on the course "boosted my confidence." He also went on a stained-glass painting course and had had a couple of pieces exhibited in a local gallery. This was news to his support worker. One of the images showed a stag, about which he said: "I did a picture of a stag. It represented me really. It was alone, on a hill and I did him bellowing out, it's not really a cry for help but it's in that sort of vein" (Figure 4.1).

C showed the digital story at the GEM at a Christmas lunch and the support worker contacted me afterwards to say "C let us play the film at our volunteer lunch. Everybody loved it. It was the most talked about thing afterwards. Thank you so much." The other participant, V, had spoken about volunteering at a Stroud charity, which had then turned into a part-time job. For them, it was knowing that there was going to be some longevity to her work that made her feel assured "that's great for me to know that it is ongoing" and she ends, "I feel excited because I can put proper jobs on my CV." The third participant wanted to share his story with young adults and teenagers. The biggest change in his life is being a father, and he has custody of his son for four nights of the week. He says that he did not have the "best upbringing" and he didn't "look at my parents

---

[13] www.mysty.eu

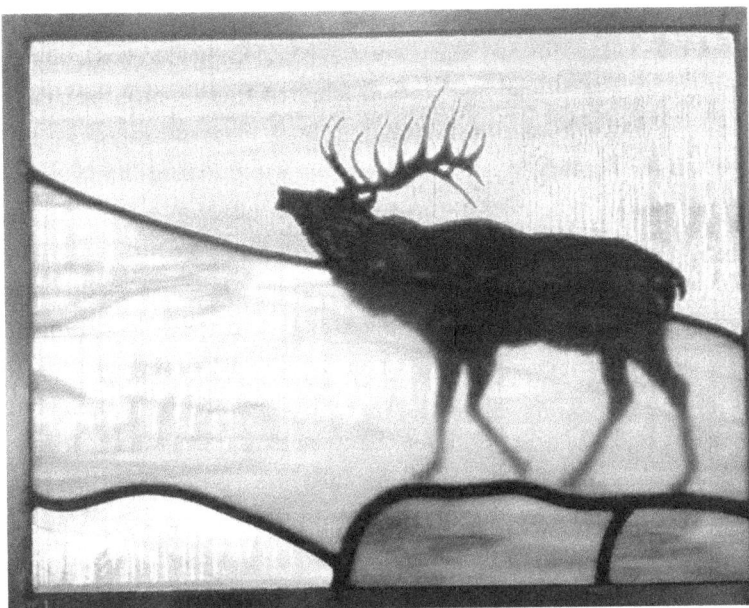

Figure 4.1 C's Stained-glass Stag. ©University of Gloucestershire.

as people I couldn't rely on," he was never able to be "top dog" and he hated his parents for that. He talks about misery and depression acknowledging that it's "so hard to explain it properly" and that "I knew, as I grew up, that I had to do things different." He talks about his love of growing things, and of starting to grow vegetables for his son, and then some chickens (Lilly and Tilly). J talks about the importance of seeing his son eat food that he has grown himself, that he has respect for the animals producing the eggs, nourishment that his parents never gave him. Halfway through the five-minute narrative, he switches tack to talk about how he got into "a bit of trouble" the previous year, which resulted in him losing his license, which impacted his ability to find work. He admits that he was in a bad place and wants his digital story to be shown to kids in school so that they would not do what he has done.

The themes of change, self-growth, security, and self-esteem emerged across the three stories, all produced because something had been accomplished through the GEM project and this was now ready to be listened to in order to gauge impact. The listening part of the project was therefore a kind of verification process, one which ensured that the project was "working" and able to be filtered up to managers and the funding bodies. It worked to help people into work, to get them back on their feet. These three voices were all scarred by events that are not detailed in the stories themselves, details of which were divulged to me by their support workers. In part, the process of being listened to was testimony to their success at overcoming their pasts to achieve either a job, or the confidence to consider trying to be able to get one.

Across contemporary Western societies we use the term "unemployed" to categorize millions of people whose lives operate outside of work. The listening exchange that went on in these three cases, went someway to spotlighting the histories and complexities that lie within that term. Of course, all the work sits on the same foundation, that is that a neo-liberal capitalist model of work is aspirational, that the good citizen is the earning citizen.

The last one I want to discuss is that of G, with whom I worked in June 2022 on the same project. G now works in a charity called ArtSpace in Gloucester, and wanted, according to the support worker, to tell his GEM story. He had a history of substance abuse and had been the main carer to his cancer-afflicted father. During his father's (terminal) illness, G had promised to change his ways, try to get clean, and make a "better life" for himself. In advance of our meeting, he had sent me a number of images that he wanted to talk about and when we met, we discussed how to organize them so there might be a clear narrative. He hadn't sent me a photo of a key ring which said, "Keep Calm, George" (his name) with a crown motif at the top, a play on the popular British slogan that comes from a Second World War poster and enjoyed currency in different mutations and products in the 2000s), and with some encouragement from his support worker told me about his father. We decided to start his story with this, then went on to discuss some events, flying in a small plane from the local airport, traveling to a Premier Inn in a different part of the country for the first time, being a volunteer at a local historical site and being a clothed life model for the painters at ArtSpace.

G was reluctant to maintain eye contact with me but digital storytelling methods of co-looking or listening, enabled him to talk in detail about the images we were both looking at. This enabled G (as it does for all digital storytelling participants) the space to reflect on the image without having to be engaged with the person recording; it is almost like talking to someone in a car, where the lack of direct eye contact has some kind of freeing up effect. He was sharing his comments with me, I was more of a bystander than an interrogator/interviewer (De La Fuente 2019).

G's story was very long, around twelve minutes, and when I went to edit it in Premiere Pro, I added a few of my own stock photographs (of hotel beds and of maps) to break up the images and provide some visual stimulation. After sharing the first edit with G, he replied with some more photos of the hotel they had stayed in and asked for me to switch those in for the ones I had done. After doing this, we again shared the draft edit and he gave me the GEM logo to add to the final slide. This was the first time a participant had been active in the editing process and was very much an example of a co-production. G kept a copy of the final version of the story, as did their support worker.

I mentioned in the methods chapter, that there are ties that bind a researcher to a funding body, expectations, and criteria to be met. Sometimes the final payment is reliant on these fulfilments, as is the case with Erasmus+ projects. Work that was outlined in a bid needs to be completed. The GEM project is an attempt to get people out of unemployment, to enable them to upskill, train, and be more confident. These are all laudable and it would be odd if a support worker had contacted me to ask me to produce a story with a participant for whom GEM had been an utter disaster. It became apparent that these listening moments were part of a process of verification; that is, the GEM participants' stories were testimonies about moving toward acceptable citizenship. Berlant calls this "a mode of belonging . . . that circulates through and around the political in formal and informal ways, with an affective, emotional, economic, and juridical force that is at once clarifying and diffuse" (2011: 230). There was a force evident in the work with GEM, it was effective; it got results, people got work, they stopped substance abuse, they built up their confidence. The time afforded to them through listening to their stories verified those outcomes, enabling the GEM management to gauge the efficacy of the project. Sharing the stories among GEM participants had been positive; voices were heard. This convivial listening was about moving people into the kind of ambient citizenship that Berlant outlines in terms of a "belonging." GEM's aim is to move people into work or training. The listening goal of all the stories was to hear the journeys into viable citizenry and away from "problematic" pasts dogged by substance abuse, criminality, or mental health issues.

Far from belittling that, the digital storytelling process was an affective listening point on these journeys and enabled a degree of critical as well as emotional recognition of the changes some participants had had to make in order to progress. It also, in parallel, cements the notion of the acceptable citizen as the working citizen, ordinariness via employment. It also operates without recourse to critiquing larger structural inequities in relation to education, health, and housing. The sonic agency (LaBelle 2018) that was afforded the GEM participants has been crucial to their self-esteem and this is happening within a macro neo-liberal economy that lauds and lumbers the individual "ordinary" citizen.

Not Listening

In a "case for abolition" of borders, Bradley and De Noronha spell out how "Citizens are the included, the members; the nation is a fraternity a family and a people. . . . This is why, when minority citizens (usually Muslims) adopt 'anti-national' positions, they can be stripped of membership and subjected to exclusion, expulsion and death" (2022: 98). The process of "stripping" starts with not listening. There are some voices that are positioned as "noisy" and unwelcome. In contemporary Britain, they belong to political protestors, notably those fighting climate change and to small groups of people whose activities within urban contexts are deemed too disruptive to be listened to. "Noise is that which exists beyond our control" (Thompson 2017: 22) and these noisome bodies are positioned as outside the ordinary; too loud to be counted as model citizens. They are part of a "politics of struggle around who speaks and who is silenced, which actors are heard, and which are ignored, which voices actually resist and which legitimize the border" (Chouliaraki and Georgiou 2022: 149). Chouliaraki and Georgiou write about symbolic, digital, technological, and palpable borders in relation to migration, and I work some of their ideas through to conceptualize belonging and place in the next chapter. Here, I use their idea of the border between acceptable citizen and problematic one (2022: 7, 12), and think about the non-listening that is constitutive of that border. Some voices are deemed too noisy for inclusion. These are voices on the edges of citizenship, whose sonic existence is circumscribed by law and ordering. As Berlant writes, "some noise is sanctioned and invited to dilate, while some noise calls out the police" (Berlant 2011: 230).

On June 28, 2022, one such voice belonged to SteveBray. Known as "Stop Brexit Man," Bray was a vocal campaigner against the UK's departure from the European Union and stood outside the Houses of Parliament in Westminster every day, vocalizing his protest with the use of a megaphone. He had been "being noisy" for over six years. But on June 28, 2022, in line with new legislation passed by the Police, Crime, Sentencing and Courts Bill 2022, police confiscated his amplifiers,[14] because he was forbidden from carrying out a "noisy protest" in a designated area outside Parliament.[15] This new legislation was the British government's Police, Crime, Sentencing and Courts Bill, which had come into

---

[14] https://news.sky.com/story/police-seize-amplifiers-from-stop-brexit-man-steve-bray-outside-parliament-12642026, https://www.itv.com/news/london/2022-06-28/police-seize-stop-brexit-protester-steve-brays-amplifiers-outside-parliament
[15] https://www.lbc.co.uk/news/steve-bray-stop-brexit-man-sadiq-khan/

force on April 28, 2022, after much debate in both the House of Commons and House of Lords. The Home Office website declares that:

> The Act will allow police to place conditions on public processions, public assemblies and one-person protests where it is reasonably believed that the noise they generated may result in serious disruption to the activities of an organisation carried on in the vicinity or have a significant impact on people in the vicinity of the protest.[16]

It goes on to specify how "This measure has nothing to do with the content of the noise generated by a protest, just the level of the noise" and its impact, on who, for how long and how intense that might be. One iteration of the bill included the phrase "serious unease" in relation to the perceived impact that noise might have on listeners, before the House of Lords insisted on it being removed.[17] Commenting on its ambiguity, Lord Coaker, Labour Party Lord, argues:

> My Motion C1 maintains our previous position that the noise trigger should be removed in full. Our Amendments 73 and 87 remove the Government's proposed noise trigger, which would allow the police to put conditions on marches or one-person protests which get not "noisy" but "too noisy." The Government have still not made the case that the power is proportionate, and the more we ask, the less they seem to know about how it could possibly work in practice. For example, the government Amendments 73C and 87H on "serious unease" show that the Government are still in a hole and still digging in recognising that there are problems with the definition of what "too noisy" means.[18]

Bray was an audiovisual fixture in British politics, and the seizure of his speakers by the police had provoked reactions from across the political spectrum. On June 22, 2020, on LBC radio, the Mayor of London, Sadiq Khan, noted that while some might find his protest annoying, it was problematic to consider it criminal.[19] At 2.18 p.m. on June 22, 2022, the same day, a right-wing Conservative Party politician, Andrea Leadsom, posted on Twitter that:

> Steve Bray has spent six years screaming abuse through a loudhailer at me and many others as often as he saw us for the "crime" of trying to fulfil the democratic

---

[16] https://www.gov.uk/government/publications/police-crime-sentencing-and-courts-bill-2021-factsheets/police-crime-sentencing-and-courts-bill-2021-protest-powers-factsheet
[17] https://bills.parliament.uk/publications/46242/documents/1746
[18] https://hansard.parliament.uk/lords/2022-03-31/debates/5B3AFEB0-0ACC-4980-921A-8A92B7E6166C/PoliceCrimeSentencingAndCourtsBill
[19] https://www.lbc.co.uk/news/steve-bray-stop-brexit-man-sadiq-khan/

decision of the UK to leave the EU. This action by the police to stop his violent protest is very welcome.[20]

Leadsom's social media post equates noisy vocal protest ("screaming") to violence. Little did she know perhaps how close to Attali's idea of noise she was: "*noise is violence*: it disturbs. To make noise is to interrupt a transmission, to disconnect, to kill. It is a simulacrum of murder" (2003: 27). For Leadsom, the oppositional scream can now be silenced.

It was not the first time that a Conservative government had policed sound, it has a record of aiming to curtail noise in relation to place and parties. In 1994, the Criminal Justice Bill was introduced in a bid to stem the free party scene associated with Rave culture and its famous line equating noise with repetitive beats:

> This section applies to a gathering on land in the open air of [F120] or more persons (whether or not trespassers) at which amplified music is played during the night (with or without intermissions) and is such as, by reason of its loudness and duration and the time at which it is played, is likely to cause serious distress to the inhabitants of the locality; and for this purpose—
>
> (a) such a gathering continues during intermissions in the music and, where the gathering extends over several days, throughout the period during which amplified music is played at night (with or without intermissions); and
>
> (b) "music" includes sounds wholly or predominantly characterised by the emission of a succession of repetitive beats.[21]

More recently, there have been attempts to police specific sections of urban youth, notably young Black men (see Fatsis 2019). Youth has always been "trouble," either too noisy (Hebdige 1979) or in the wrong place (Deller 2018), or both. And Black youth more so, and policing its music has a long history (Hall 1978[2019]; Cloonan and Street 1997). This latest iteration of dampening sound is directed at voices, not music. It is the sound of dissent which is being turned down. And this dissent is largely in relation to climate change, and groups such as "Extinction Rebellion," which, like Greenpeace before it, advocate direct action.

Gallagher asks us to "Consider how a space might 'vibrate' after a loud retort has echoed, or a street might still hold the sonic memory of a recently passed demonstration" (2017: 627) and his suggestion goes some way to offering a route

---

[20] https://twitter.com/andrealeadsom/status/1541772945042931714
[21] https://www.legislation.gov.uk/ukpga/1994/33/section/63

into thinking about what this bill is about, at its core. It is an acknowledgment that listening matters, that noise that is encountered is affective and is memorable. It understands the "connected" part of connected listening; that is, how sound is imbricated with belonging and memory. Things that are too loud leave imprints; they leave traces. They are testimony to dissent and to opposition. Marchers make noise in order to disrupt, and I go back again to Hebdige's (1979) brilliant sentence about the upset that subcultural groups do to society, that is they "interfere with the orderly sequence of events" (p. 90). From the Jarrow March in 1936 to the anti-Brexit marches before the EU Referendum in 2016, from the Miners protest and battles at Orgreave in 1984 to the Poll Tax Riots in 1991, protest has used sound to literally voice its concerns. It is *supposed* to not be ordinary. It uses the street to shout in, not to shop or drive through. It is a vocal demand, "listen to us."

In a paper on noise, morality, and citizenship in Australia, Rosenberg sums up the "moral geography of noise" (Rosenberg 2016: 190), calling it "a site of conflict which involves debates about moral and ethical behavior and what it means to be a good citizen" (2016: 190). His argument pivots on the notion of a "residential ethics," where there are links between noise, citizenship, and care, which is demonstrated through the minimizing of noise and so "this moral geography of noise, about who makes noise where, leads to proclamations about appropriate social behavior, where quiet residents are good, civil citizens, and noisy residents bad and uncivilized" (2016: 191).

News outlets in the UK reported on such a case in May 2022 (Taylor 2022). It centered around a 73-year-old man, named Ernest Theophile, his fellow dominoes players, and a square near the junction between Harrow Road, Westbourne Park, and Maida Vale in London where noise went beyond the agreed "clamour of urban life" (Back 2007: 118). The news report went as follows:

> A dominoes player has won a court case over a ban against him and his friends playing the game "loudly" that he said was racially motivated.
>
> Ernest Theophile took Westminster city council to court after it granted an injunction banning social gatherings in Maida Hill Market Square in north-west London.
>
> Theophile and his friends have been gathering in the square for 12 years, to chat, socialise, and play dominoes, cards and backgammon. However, the council banned them from congregating there in January 2021, citing noise and antisocial behaviour problems. It said it had received more than 200 complaints.

However, the 74-year-old took the council to court, saying its order was racist as it discriminated against Caribbean culture. Theophile's family arrived from Dominica in the 1950s as part of the Windrush generation. (Taylor 2022)

This report details how community services for older people in that area of London, one that experiences severe deprivation, have been severely curtailed, and that there is nowhere else for him and his peers to congregate and socialize. The significant comment from Theophile is this, that "If you are West Indian, you just can't play dominoes without making a bit of noise." Theophile's family came from Dominica in the 1950s. He is part of the Windrush generation.

Theophile punctured the skin of contemporary British notions of citizenship by being noisy, an immigrant, and old. There he and his friends were, outside, in company, playing a game; all activities that might very well be the opposite of the home-owning, nuclear familied complainants, of whom there were 200. He brought the empire right back into that London square, clacking the dominoes down as he and his elderly friends sought to carve out their own public and playful sphere. He was not doing his citizenship right, to borrow from Butler's famous (1990) phrase (on gender). Not only that, but he was not doing his age right. He and his friends were out of place. They were blurring the boundaries of in and out, public, and private, Dominica and the UK, their noise was seeping out from those pasts into a present that could not accommodate, could not listen to, those sounds. These sounds came from "the margins of those dominant articulations of modernity, black diasporic sound cultures [which] have continued to provide alternative playlists for the modern condition" (James 2021: 5). The lawsuit against them was one wrapped up in a fear of the abject, of Otherness creeping into their immediate sonic environment, unwanted sounds from other places. We are back to Adorno, and I wish we weren't. Back to his comments on Blackness and noise, as Thompson notes in her work on "unwanted sound" where she writes:

> There is a duality to the noise of the social "other." On one hand, the dismissal of particular bodies as noisemakers demeans and trivializes—it asserts the inferiority of another who is incapable of meaningful or pertinent comment. To be "mere" noise is to be worthless, incomprehensible, extraneous, ugly or unpleasant. Yet noise also carries with it the threat of disorder or disruption and is thus wanted to the ears of the establishment. (Thompson 2017: 28)

He and his friends' sonic activities indicated that there was some kind of "sonic color line" (Stoever 2016) that had been transgressed. This line "describes the process of racializing sound" and the click clack of dominoes being slammed onto tables indicating Blackness where Blackness is noise (Thompson 2017; Radano and Olaniyan 2016: 8), and noise is unwelcomed and not a component of the docile citizen. Listening is "deeply enculturated. Through a cultural process that divides signifying vocal sounds from non-signifying vocal sounds. We learn to value each differently" (Eidsheim 2015: 100). Or we devalue them and seek to silence them.

Coming from a different angle, concerned more with grounded experience than the philosophy of noise and affect, Western's work on sound and migration offers a succinct history of the relationship between noise and race, which he argues are indexed (2020: 300) and uses Radano and Olaniyan's (2016) work to argue that noise is a "concept employed by European colonialists to domesticate the sonic expressions of those subject to imperial order" (2021: 300) and "in the archives of colonialism. Noise often serves to differentiate colonizer from colonized, civil culture from barbarous nature, human from non-human" (Thompson 2017: 28). This is what was happening to Theophile in London in 2022. He and his friends' dominoes were producing sounds that were "a productive and performative force that creates space" (Gallagher 2017: 619), and that space was a sonic slice of the empire talking back. And the emperor didn't want to listen. It was a border in the middle of London where a battle for "spatial control" (Kanngieser 2013) through listening was happening.

In a work on "violent borders," about the current state of migration, Jones writes, "There is a powerful idea in the media and in wealthy societies that violence at borders is inevitable when less developed, less orderly countries rub against the rich, developed states of the world" (2017: 4). London's Maida Hill market square was, in this instance, a border where the sounds of the "less orderly" (and also, less "ordinary") rubbed up against the ears of those who did not want to listen, to let in. For these ears, the dominoes' slapping was a "calamity" (Radano and Olaniyan 2016) because it asked them to "be with" (Bull in Erlman 2004: 188) Theophile and his friends. It was too close.

Listening (and not listening) is also about being exposed to "stuff" that is not just too "close for comfort" but which is threatening. Western (2020) writes about the history of the demarcation of noise as pollution: "noise in Euro-modernity accrues metaphors of pollution and foreignness: something to be regulated, abated, eliminated" (e.g., Bijsterveld 2008: 300). I would link up what Western is arguing about how noise is controlled, with what feminist writers such as Kristeva (1982) Creed (1993), and Grosz (1994) have done with the concept of the abject. Abject is about a throwing away of matter that used to be part of you/us, but is no longer; blood, semen, the corpse, housed in rubbish bins and cemeteries (Gardner 2015: 118). Kristeva uses the return of that matter to understand how horror works and this, coupled with its other adjectival meaning of awful, along with Kristeva's use of it as "that which disturbs identity, system order" (1982: 4) has been used to marshal how women who were leaky, old, ill, noisy, troublesome, and so on were silenced. There are parallels with this that can be redeployed to think about what was going on in London in 2022. Because sound is "a force that disrupts and reworks common spatial concepts such as boundary, territory, place, scale, and landscape" (Gallagher 2017: 620), it can start to work in the same way that other beings, objects, and people do that are abject, where the abject is that which is disavowed.

In his work on sound and citizenship in Athens, Western notes how prioritizing sound in narratives of displacement foregrounds the "messiness of life across borders" (2020: 296). In the opening of his book about space, place, and Hip Hop, Murray Forman (2002) described the sonic experience of driving into cities in the States; about how the radio changes and the listening of it makes you aware that you are in a new space. This happens too in London, when you drive up the A40 or M4, you can turn your dial, like Forman did, to pirate and official radio stations playing Grime, Hip Hop, and Bhangra Radio has always had this characteristic, which is continued by the BBC local radio stations, each with their limited broadcasting range and difference is demarcated through the airwaves (Keogh 2010). What seems to have happened with Theophile and his dominoes is that they breached an unknown barrier and came within range of listeners who had not chosen to hear him, and his friends play dominoes, nor appreciate that living in a densely populated urban area like London might expose you to sounds. These temporary apertures that offer moments of connected listening are resisted, as the court case revealed. These were Raced sounds (Stoever 2016: 21) (dominoes and Dominica) bursting into spaces whose sonic economy consisted of traffic and commerce. It did not want to listen to the sounds of Theophile and his friends. They had no permit; they had no permission. But sound has "miasmic qualities or what geographers have called affective atmospheres (Bissell 2010)" (Gallagher 2017: 625), and as such, seeps across, into, and beyond. We come back to Berlant's idea of the ambient citizen, the moving citizen, and questions around "whose noise matters, whose immediacy-pressures rule the tendency of the situation—and who controls the zoning" (Berlant 2011: 230).

Conclusion

Belonging is yearning and loss, the "might be" and the "has been"; always promised, often grieved over, always constructed through the lure of the ordinary, whose sounds do not disturb. Convivial listening cannot operate when conviviality is denied. Some listening is performatively dislocating, it renders its subjects Other, casts them out. These voices are fractures. They are not granted an audience (Chouliaraki and Georgiou 2022: 152 citing Dreher 2009: 99). The cultural citizen is produced by affective regimes that flow through sound, across the airwaves. *Listen with Mother* and *The Listening Project* were both institutional attempts to shepherd into citizenship (be ordinary, we can archive your ordinariness). And GEM is both a positive space for engaged listening and a model of productive citizenry in action. Operating a sonic safe space, the edges are where the frays and strays accumulate, the noisy protestors and first-generation immigrants. Disallowed citizenship. Be quiet, you who are too noisy, too subaltern, too agitating. The sounds of empire and opposition split the sonic security of a policed sonic state. Noise frays the edges of this polis where inclusion is determined by a recourse to an imaginary intimacy and through the lure of the ordinary.

5

Listening

Migration, Voice, and Place

On the evening of June 14, 2002, a chartered plane meant to be flying from an RAF military base in Wiltshire, England, was grounded. Its destination was Rwanda, and its passengers were, depending on which newspaper you read, "illegal immigrants" (The Daily Mail), "refugees" (Manchester Evening News),[1] or "asylum seekers" (BBC and SKY News[2]). Frantic negotiations between lawyers for the seven[3] people who had been randomly selected to be flown to the central African country's capital, Kigali, took place after the European Court of Human Rights issued an injunction for one of the people on board. Over the course of the next few hours, as lawyers argued each case, the passengers were taken, one by one, from the plane, until it was empty.

This event was the first manifestation of the British Government's agreement with Rwanda to "offshore" refugees, by sending them "back to Africa" in a grotesque and grandiose political gesture of exclusion beyond the boundaries of the nation-state. It illustrates some of the ongoing media discourses around migration into the UK that are about "western econom[ies] of voice" (Chouliaraki and Georgiou 2022), power, and silence. What was very apparent from the coverage of the incident was how noisy the press was about it, both those in favor and those who were vehemently against it, and how silent those on the plane were. Not only did we not see them, the "invisibles" (LaBelle 2018: 39), but we didn't hear them. Silence can sometimes be a weapon of protest (Lykkes, Bianco and Távara 2020) but here, it was used to depersonalize and erase (Voegelin

---

[1] https://www.manchestereveningnews.co.uk/news/greater-manchester-news/plane-suspected-flying-refugees-rwanda-24227555
[2] https://www.bbc.co.uk/news/uk-61806383, https://news.sky.com/story/rwanda-deportations-one-asylum-seeker-loses-legal-challenge-against-removal-to-african-nation-12633715
[3] https://www.independent.co.uk/news/uk/politics/rwanda-deportation-flight-home-office-b2099601.html

2019: 114) and once again deny agency to bodies marked as migrant (LaBelle 2018: 6). These voiceless and unseen "passengers" were grouped together, their "individual histories and personalities" erased (Nyers 1999 in Chouliaraki and Georgiou 2022: 108). They were generalized within a reportage discourse that did not name them; generalized and so depersonalized (2022: 117). Language is a "technology of power" (Chouliaraki and Georgiou 2022: 106) so to be prescribed within it as "migrant" with the history and connotations that term brings with it, together with not having access to that technology, not being able to speak.

The chapter is about listening beyond the statistics (Chouliaraki and Georgiou 2022: 108) to people whose names we got to know and whose stories we heard. These stories are part of Mapping the Music of Migration, an Erasmus+ funded European-wide project. Through a series of training workshops, manuals, and apps, it established connected listening moments to generate migrant-led stories aimed at facilitating multimodal and open-ended moments of exchange, or "modes of engaged attention," "creative resistances and acts of radical sharing" (LaBelle 2018: 7). The project was an attempt to counter the "bureaucratic and institutionalized practice of exclusion, containment and marginalization" that arguably characterizes migration (Krüger and Trandafoiu 2014: 16), a "pressing" issue (Voegelin 2019: 80). We wanted to create a "new contact zone, the listening event" (Kheshti 2015: 22), to have people speak and be heard, in their own voices, and to have those listened to not only for what they were saying, but how they resonated: "Anthropologists and forced migration scholars tend to focus on voice as a metaphor for agency, forgetting that voices also do a lot of work as sounding entities. Voices are sonic phenomena" (Western 2020: 304). MaMuMi aimed to add sound where there was silence and offered specificity and ordinariness to counteract prevailing discourses that render the migrant homogenous and either a threat or in need of saving (Braidotti 2013; Chouliaraki and Georgiou 2022). We wanted to de-exceptionalize displacement and provide routes to understanding through listening (Gardner and Hansen 2023).

I have called the chapter "Listening: Migration, Voice, and Place" because there are tensions between fixity (place) and flux (migration) that are narrated in some of the stories we recorded. The migrant is a figure who is *always* determined by movement or journey. Migrants are by definition, *moving people*. This is important because it is their fluidity as well as their Otherness that dictates their disposability, indeed, it is this fluidity, this status as moving beings that renders them Other (Mbemebe and Meintjes 2003: 2). Other bodies who have moved through and been lauded for it are marked by whiteness and authenticity. The migrant is different to the (male) flâneur who moves at volition through the bourgeois city, and whose travel is fueled by wealth and privilege. The migrant is different to the "nomad," a recognizable authentic (Fraser 2018: 28), moving through space on the back of tradition and volition. No, the migrant is circumscribed by movement, forever to be determined by that word, that journey. To be a migrant is to be liminal, between places and forever refused full entry into the new or back to the old.

Fraser argues that "all acts of migrancy are far more complicated than such a list of arrivists would suggest" (Fraser 2018: 28) and it was this complex space that we were looking to open up. In 1996, Bauman reflected on the similarities between tourism and migration noting that "Like the tourist, the migrant is in place, but not 'of the place'" (Bauman 1996: 29 in Kruger and Trandafiou 2014: 1). This assertion harks back to the previous chapter in belonging and citizenship, where fluid or marginal bodies are not deemed worthy of inclusion, and the word "migrant" is one that has lurking in it the shadows of forced travel and uncertain receptions. Its suggested flux is a quality that LaBelle argues is characteristic of sound, since they are both determined by movement across bodies and borders, or as LaBelle writes sound moves through "matters and bodies [and] is deeply linked to expressions of migration and transience" (2018: 19). Putting the two together to enable some new soundings was one of the motives behind the project. Using ideas on voice and silence (Voegelin 2019; Chouliaraki and Georgiou 2022), sonic agency (LaBelle 2018), displaced listening (Western 2020), and mapping (Sweers and Ross 2020), MaMuMi was an attempt to counteract dominant narratives around this figure of "the migrant" by "adding volume" to people with a migration history. As such, when it comes to reading and listening to the voices of the project itself, I have left many of their voices in, so they can sound.

Mapping the Music of Migration

"Mapping the Music of Migration," known as MaMuMi, was a two-year Erasmus+ funded research project led by the University of Gloucestershire (November 1, 2019 to October 31, 2021) with a grant of €245,297.[4] It centered on migrants

---

[4] https://erasmus-plus.ec.europa.eu/projects/search/details/2019-1-UK01-KA204-061966

telling stories about songs and was developed in response to a call from the funding body partly in response to the increased migration into Europe over the past decade, to the "increasing presence of bodies out of place ... [that has] taken on pronounced intensity and number within the contemporary global environment" (LaBelle 2018: 91). The Greek partner KMOP articulated it clearly on their website:

Challenge
The ongoing context of rapid migration into Europe and the politics require an innovative response in order to counteract current fears and open a more personal perspective to the life of migrants.

Migration has presented various challenges across different national contexts and is a transnational phenomenon. It is important that a transnational solution is developed and for wider impact to be achieved.

Innovation
By focusing on telling stories about migrant music and mapping those stories and sounds into an interactive map, **MaMuMi** aims at combating discrimination and promoting intercultural competences to support trainers working with migrants and refugees.[5]

The project team known as a "consortium" included non-governmental organization staff and academics from Bulgaria, Cyprus, Greece, Italy, Norway, Spain, and the UK. There were two academics from the University of Gloucestershire and the Inland Norway University of Applied Sciences along with five non-governmental organizations: Know and Can (Bulgaria), CSI (Cyprus), KMOP (Greece), CSC Danilo Dolci (Italy), and Caminos (Spain). These third-sector organizations worked in the fields of development, education, and arts, and had years of experience working with migrants directly and with migration support groups and government agencies. Many of the groups had worked with each other on previous projects, although MaMuMi was the first time they had collaborated in this formation. The Norwegian institution had not been involved in any Erasmus work before, and I had invited them onto the project because of their expertise in the field of popular music (Dr. Kai Arne Hansen) and ethnography (Dr. Camilla Kvall).

Our project statement advertised how we were "focused on talking about music and song as a tool for intercultural competency" and were "using

---

[5] https://www.kmop.gr/projects-vf/mamumi/ (accessed February 19, 2022)

storytelling about music as a positive mechanism to counter negative stereotypes and to open up enabling enunciative spaces."[6] Rosi Braidotti says that "Symbolic of the closure of the European mind is the fate of migrants, refugees and asylum-seekers who bear the brunt of racism in contemporary Europe" (2013: 52) and the Erasmus + call and our response was an effort to counterbalance that "closure," through facilitating a form of connected listening.

The inspiration for using song to drive a story came from my work with the Women, Ageing and Media research hub at the University of Gloucestershire, which, led by Professor Ros Jennings, had started to develop the idea of "Inheritance Tracks." Ros had used a similar model to a BBC Radio 4 show of the same name,[7] where a guest reflects on songs they have inherited from their parents and songs they would like to pass on. In the "Inheritance Tracks" workshops at the Women Ageing and Media summer schools (in 2016, 2017, and 2018), we asked attendees before the event to choose two such songs. The workshops were places of listening exchange where participants shared narratives triggered by these songs, and the stories of some of these workshops are collected in a book called *Troubling Inheritances* (2022), edited by Ros, Sara Cohen, and Line Grenier. The book queries the idea of inheritance and looks at the intersection of music, age, and memory outside of a therapeutic setting. I use a chapter in that book to reflect on a Song Story Workshop that I led in Palermo in October 2018, and which was the first time I had used music instead of images in a group storytelling exercise. The Workshop was held at the end of a meeting on another Erasmus+ funded project that was about media literacy for refugee, asylum seeking, and migrant women. There were ten participants: two academics, one support staff, and seven NGO workers, and what happened in it started me thinking about how talking and listening to others about music had the potential to disrupt "modalities of communication and expression" (Gardner in Cohen, Grenier, and Jennings 2022: 82). Doing the workshop, I noticed some key differences when using music rather than images, and reflected on it thus in *Troubling Inheritances*:

> Playing, listening to, and sharing music changes the affective atmosphere. There is a shift in an inheritance tracks workshop and it is like a key change. First, the mood alters (DeNora 2000, Kassabian 2013). And importantly, it alters for different participants according to their own previous exposure to the track

---

[6] www.mamumi.eu
[7] https://www.bbc.co.uk/programmes/b00zwntv

playing. Is it something they know? What might their relationship to it be? Music being played not only takes the individual who chose it on a narrative journey but invites all other group members to travel with them too. The resulting stories and conversations demonstrated an iterativity different to the worked-out scripts of digital storytelling. (Gardner in Cohen, Grenier, and Jennings 2022: 82)

Sound (and music) has the potential to produce "a highly malleable and charged relational area, modulating the social coordinates and territorial boundaries by which contact, and conversation may unfold" (LaBelle 2018: 8; see also Gallagher et al. 2017: 621) and to "affect emotions in strong and sometimes unexpected ways" (Istvandity 2019: 23). This is what Kathleen Stewart calls "sudden eruptions," as in when "Things flash up—little worlds, bad impulses, events alive with some kind of charge. Sudden eruptions are fascinating beyond all reason as if they're divining rods articulating something. But what?" (2007: 68). The Palermo workshop and the "Inheritance Tracks" events led me to consider that these "flashes" or "eruptions" might be part of an affective shift that could be explored further.

What became clear from doing the Palermo workshop was that working with song affords a degree of disturbance, upsetting the balance somehow, having something to do with "the contingency of borders, especially those of the affective self, and of place and time—of what belongs where" (Gardner 2022: 83). In a paper on haptics and EDM, Garcia talks about this kind of shaking up or disruption as the "bleed between modes of feeling" (Garcia 2015: 59) and this idea works well to account for both the affective shift that happened in Palermo, and later on in MaMuMi, where the "porosity" that listening engenders breaches a number of boundaries; corporeal, emotional, social (Erlmann 2014; Grosz 1994). Some of these "incursions" that had been encountered in Palermo, were written into the methodology and user manual that made up MaMuMi's "work packages."

MaMuMi Work Packages

All Erasmus+ projects have what is known as "work packages" or "intellectual outputs" that are produced within the project team and published both on its website, and on the Erasmus+ results platform.[8] MaMuMi had five "practical & reusable resources for the practitioners" (26/08/2022); a "methodological approach to using music as a tool for enhancing social inclusion tailored to the

---

[8] https://erasmus-plus.ec.europa.eu/projects/search/details/2019-1-UK01-KA204-061966

needs of migrants" (2019) that was based on our individual national reviews of policies with respect to migration and music, a MaMuMi Song Story workshop manual, a short training video, a mobile-friendly app and a research paper written by myself and Kai Arne Hansen. All of the resources are translated into the languages of the consortium and the app has a dropdown menu that also includes Turkish, given that Cyprus was a partner.

The "MaMuMi Methodological framework" and its individual seven national reviews concluded that there was a gap in provision for talking about music as a viable practical inclusion method, and our research indicated that arts provision, including music-related strategies, were not high up on a list of integration approaches. The English review reported that:

> As ec.europa notes, "immigration policy is often focused on two areas: preventing unauthorized migration and the illegal employment of migrants who are not permitted to work, and promoting the integration of immigrants into society." This integration remains at the level of access to housing, healthcare and, pending successful asylum status, employability. National and local organizations offer advice in these areas and EU Commission Reports such as the EU Indicators of Immigrant Integration (2013) report noted that integration was linked to housing, legal access, democratic process, economic stability. These are all fundamental to integration in the first instance and might be termed the "hard" integration approaches. MaMuMi will be using "soft" integration methods to build the sense of community between migrants and between members of host countries and migrants. (https://mamumi.eu/results/)

This was echoed by the Norwegian review:

> In the most recent version of The Government's Integration Goals (2015), a brochure published by the Ministry of Children, Equality and Social Inclusion, the words "art," "culture," and "music" are not mentioned once, which seems to concede an attitude towards music and culture as insignificant for integration purposes. (https://mamumi.eu/results/)

There was no national initiative in Spain to use music for integration and the Cypriot team reported that art and music were completely absent from central documents that described the Cypriot immigration and integration policy and the Greek partner reported that that if there was any provision, then it fell to private organizations and not the public sector and "recommended that public authorities take a more active role in the implementation of artistic and more specifically, musical events that will help immigrants share their culture with other people and raise awareness about migrant issues to the public." It was only Italy that had had some experience of using music in a broader arts integration strategy for migrants. It suggested that there was

> support for music in relation to integrating policies but there was nothing, per se, on "talking about music" sing national and European funds, NGOs and private organisations have carried out projects for the social and economic inclusion of freshly arrived migrants, second generation migrants, asylum seekers and not accompanied minors who reached the Italian coasts. In this process, music and the arts in general have been a powerful and effective tool for the social inclusions on migrants and refugees. (https://mamumi.eu/results/)

Based on these reviews, the project team set about producing a sustainable set of resources for use by NGOs, the workshop manual, video, and mobile-friendly app. However, our research was to some extent curtailed in March 2020 when the Covid-19 pandemic appeared, five months after MaMuMi started. We were unable to run first, our focus groups, and then, our Song Story workshops in person. Our listening now would have to be online. Because we had no experience of running face-to-face workshops on the project we had no "control" against which we could map the online ones. However, from running the Palermo event in 2018, I would note that in the workshops I ran, dancing was missing, the body was absent in that way. Zoom made things quieter, more subdued.

We also had to revise down our initial aim of recording seventy stories across fourteen workshops in seven countries due to Covid rules across the consortium which made recruiting participants more difficult. Thirty-six interviews were conducted, twenty-nine of which could be used on the app (the others were exempt due to bad audio quality). In Italy, Cyprus, Greece, Norway, Spain, and the UK, online recordings were enabled by Zoom or Skype and were between people who knew each other prior to Covid; mostly through their role as NGOs, and in Norway, through institutional connections. In Greece, the support of a cultural mediator and interpreter was needed as most participants were Arabic or Farsi speakers. In Bulgaria, participants made their own video and sent them to the staff at "Know and Can."

Another challenge that we encountered was balancing NGO experience with academic critique. The following set of questions are those that the team developed for use in the Song Story workshops. There were many discussions about these, largely to do with a disagreement on question 12: "For you, what

Question for participants	Please write answers here
Which is your home country?	
Why did you decide to come to the hosting country?	
For how long have you been in the hosting country?	
What do you do at the moment (profession/education/etc.)?	
How old are you?	
Do you use your mobile phone to listen to music? If yes, how often?	
What kind of music do you like?	
When and where do you listen to music?	
Is there one piece of music, or a song, that is important to you? Why?	
This important song - how does it make you feel when you hear it?	

Is there any song in particular that reminds you of an important event in your life?	
For you, what is the power of music?	
Is there anything else you would like to add?	

Figure 5.1 MaMuMi Song Story questions. ©Mapping The Music of Migration https://mamumi.eu/.

is the power of music?" The academics on the team thought this was a leading question but it stayed as the NGOs argued for it being one that many people would be able to answer (Figure 5.1).

This is from the MaMuMi resource pack (on the Erasmus+ platform and on its own website) and clarifies what the contents of the guide are and who it is for.

> This manual is a user guide mainly addressed to trainers (NGOs staff, facilitators, social workers working with migrants), guiding them through all the steps (including preparation, ethics and follow up instructions) needed to prepare

and implement a MaMuMi Song Story Workshop, one of the core elements of the MaMuMi methodology.[9]

The workshop manual also contained ice breakers and recommendations about sensitivity and impact. It recommended the use of a cultural mediator to "mitigate" any cultural clashes, and to prepare for participants getting emotional when they talked about a song and their home. There was the usual consent and release form, which guaranteed anonymity and made it very clear that this was important given the status of many "irregular" migrants across Europe. Anonymity was offered to all participants for the reasons above in our manual that the NGO team members had insisted on. Despite this, some participants were keen to name themselves, and these were the young men who were musicians in Cyprus and Norway (Song Stories 5 and 20), who welcomed the publicity that putting their music on the MaMuMi app afforded them.

Song Story Workshops were held across the consortium and twenty-nine stories were collected that were of decent audio quality to be mixed with the song/piece of music the participant was talking about. These "song stories" were then added to a virtual map, the MaMuMi app, which was built by staff and students in the computing department of the University of Gloucestershire. When the user clicks on a place, they hear the story (and clips of the song) and trace that individual's migration journey on the screen, from and to their current location. Again, adaptations had to be made in doing this. There was a rudimentary functionality to the line as it moved from where the journey had begun to where it was recorded; it took a direct line which may not replicate the routes people actually took. Students wanted to add in stopping points, add images from Google maps, but the stories were not about places on the way to where they were, in fact only one story, from a young man in Cyprus, referred to Lampedusa and this was as a place where many people drown.

The app, and all the other outputs, were translated into the partner languages and Turkish. We used recognizable design, or the "aesthetics of amateur ordinariness" (Chouliaraki and Georgiou 2022: 170) to bring the stories "closer." The map looked like the ones that come on every mobile phone, that take us from A to B via the quickest or cheapest route. And so, we subverted the ubiquitous Google Map, replacing route maps of desire and estimated travel times with stories by those whose narratives are often pre-mapped in dominant public

---

[9] https://ec.europa.eu/programmes/erasmus-plus/project-result-content/88d4742e-09d1-4689-9237-b1770106c7d2/O2_Workshop_Manual_EN.pdf

discourses. We did not include borders, choosing instead to erase those and use only towns and cities as points of reference for the journey to be mapped across.

Borders/Maps and Apps

Borders are places of violence (Brand 2006; Jones 2017; Walia 2021). They determine difference and are both non-negotiable and highly contingent, at risk of fracture (Gardner and Jennings 2020). They are the shifting, surveilled, yet visible places where migration is trouble(d), and which migrants are seen as "breaching."

Bradley and De Noronha (2022) argue that they are not only "at the edges of national territory, in airports, or at border walls." In fact, borders are in the everyday and everywhere (2022: 5), that they "reproduced racial and colonial inequalities . . . [but] are most intense precisely at the borders between the developed and underdeveloped world: at the edges of Europe" (2022: 9). We wanted the MaMuMi app user to move across these edges, from Africa, Asia, and Europe without encountering those border lines. This is not to discount the ongoing force that borders have, how they are manifold and manifest in tangible, visible, and digital forms (Western 2020; Walia 2021; Chouliaraki and Georgiou 2022) and that "transnational connections [. . .] produce and police borders, and the power structures that make refugees in the first place" (Western 2020: 303). These borders may also be emotionally and cognitively internalized (Cockburn 2004: 1), subject to multifarious layers of policing and surveillance. But our lines of travel were individual narratives spliced with songs; song stories with "time-slices" (Voegelin 2019: 81) fracturing back and forward, crossing and re-crossing those borders as the stories were told.

The MaMuMi app was an experiment in thinking about mapping without borders and with different connections, where perhaps we might engineer spaces of intimacy (James 2021) through listening to voices and songs recorded by the people who have made their journeys, and crossed geographical and taxonomic borders, from citizen to migrant. We, the team, had agreed that we wanted users to listen to a voice, interwoven with a song, and slowly track that person's journey from where they started out to where they are now, and travel across towns, cities, and seas without knowing where the border might be. Although clearly, their journeys to their current location would have involved crossing and negotiating borders, none of the Song Stories mentioned them.

They had been erased. We were not wishing them away, as some have criticized our project for, we were thinking how our stories simply excluded them, that they had no significance in recollections. We do not want our app to replicate the contemporary map whose borders, notable in West and North Africa and South Asia, were lines of brutality and exclusion; colonialist mappings meant to split things apart (Walia 2021). We wanted our project to make new affective formations "conducive to empathy and compassion" (LaBelle 2018: 4) built on listening. While we were working with displaced people, the app was a chance to open up new lines of connections that may, or may not, focus on place.

When the project team came up with the idea of "mapping," using it as our title, it was an attempt to make visible "varied migratory journeys, musical experiences, and affective memories—as well as for probing at the relations between these. By mapping project participants" "song stories" onto an interactive app, the project aimed to spotlight cultural diversity and intangible cultural heritages (Gardner and Hansen, forthcoming; Sweers and Ross 2020: 175–6). This version of mapping, which, in its UNESCO form, is allied to the word "cultural," becomes an arbiter of institutionalized discourses around diversity. Sweers and Ross note how "Cultural mapping is regarded by UNESCO as one of the most central methodological resources for safeguarding cultural diversity" (Sweers and Ross 2020: 13). This is an archival move, "a way of defining what culture means to the community, identifying the elements of culture that add value . . . recording, preserving or building on these elements in new and creative ways" (2020: 14). This mapping is like an audit, but it can also "reveal a so-called third space" (2020: 16, they refer to Kjeldsen 2020: 37). This term has psychoanalytical overtones (Winnicott) which might not have been initiated but which I want to literally "play" with here. Our Third Space in MaMuMi was the space of affective listening and storytelling. We wanted to map that space by thinking about the different lines across which music and migrants traveled and how they coalesced together in their song stories.

Listening and Lines of Familiarity

The basic premise of our workshops was that they provided opportunities for listening exchanges (LaBelle 2018: 4), "encounters" (Western 2020: 303) and "creative engagements" (Western 2020: 294) and between October 2020 and July 2021 we recorded thirty-six Song Stories. Seven could not be included;

either interviewers were talking at the same time, or the sound quality was not good enough for reproduction onto the app. Three participants sang their own songs, two of whom were musicians, recorded in Cyprus and in Norway, who produced and arranged their material themselves. A young French/Breton man sang an old Breton folk song unaccompanied and recorded himself singing it. The stories were recorded by members of the seven teams, all with different relationships to the people they were talking to and with the music these people chose to talk about. Through discussions, the team was clear that it was not for the workshop facilitators to direct the story or to edit it in any way. Our job was to listen and record. We did not want "carefully curated narratives with predetermined storylines" (Fernandes 2017: 2) and our methods for the recording and transcribing were clearly set out in the manual. The majority of the facilitators were project team members but for KMOP, which is a large organization, staff were at one removed from the MaMuMi team and were working from the Song Story manual, recording online. The stories are the results of these temporary liaisons. They are snapshots: moments of collision between these three; the participant, the recorder, and the song, which are all are particular to that moment and to each other in that moment. Some are driven by memory, not in the way that Istvandity explored in relation to lifetime soundtracks, but more in keeping with the *Troubling Inheritances* methods, where one moment reveals the dynamic relationships across self and others, past and present, this "eruption" or "flash" (Stewart 2007: 68). Istvandity argues that "musical memories [can be thought of] as networks that are constantly in flux" (2019: 14) and we were recording one moment, the colliding point, which became the song story, some of which specifically referred to the past. Songs provided lines of identification that took people "back" to places that they were no longer part of. These places may be, as Auge in his work on non-places, "place of identity, of relations, of history . . . a system of possibilities, prescriptions and interdicts whose content is both spatial and social" (2008: 43). And so, there are stories about gardens and mountains, of people left behind, rituals remembered and vital WhatsApp groups. The following table shows where the story was recorded, what song or piece of music was used and where that too, was recorded. I had added this information to think about types or lineages of listening that came through participants' choices. Neither myself nor my team followed this up, but it is another potential strand to explore when doing work like this. What it does do is give a variety of music that is sung in Arabic, Farsi, Urdu, Greek, German,

Italian, Ukrainian, Breton, and English. I would like to think it suggests something about "families" of popular music that revolve and are accessed by listeners; West African participants chose Kora music, Arabic speakers chose music sung in Arabic, Anglo-American pop music was not dominant in any way. This is not something that we, as researchers, fully explored during the lifetime of the project, but the choices make for some reflective thinking that could reference popular music and "aesthetic cosmopolitanism" (Regev 2013), popular music cultures in the Middle East and North Africa (Hashemi 2017; Davis 2004; Davies and Bentahila 2006), and German Schlager (Mendívil, Jacke and Ahlers 2017) as these forms of musical genre were appearing in the song choices. The last column is a marker of the overall theme of the story that I have added, trying at the most basic level to codify the songs thematically. This is crude, but it does "key word" them in a way that is useful. There are only three specifically about the journey that they took from where they spent their childhood to where they were in 2020, five are about "home," and there are three stories about grandparents. Family is key to these stories, and as Stewart writes, the relationship between narrative and family is crucial: "her family, like all families, built its skin around dramas and luminous little tales with shiny scenes and vibrant characters" (2007: 16). I want to carry that analogy forward, that stories are the skin-making mechanisms for identity building and that, when fused with song, via reflection and production, they become affective motors that can indeed, bleed through places and terms.

	Where participant's song story recorded	Participant's nationality	Musician's nationality	Chosen track/title and artist	Song story focus
1	Sofia, Bulgaria	Brittany/France	French/Breton	"An Hini A Garan," Denez Prigent	Home
2	Cyprus	Greece	Greek	Etizmao	Overcoming difficulty
3	Cyprus	Gambia	Malian	"Lampedusa," Toumani and Sidiki Diabete	Difficult migration journey
4	Nicosia, Cyprus	Gambia	Gambian	"Tamala," Foday Musa Suso	Travel and xenophobia
5	Cyprus	Cameroon	Cameroonian	Daviny	His journey and God
6	Cyprus	The Netherlands	North American	"I Will Survive," Gloria Gaynor	Overcoming difficulty
7	Athens, Greece	Afghanistan	British (Iranian born)	"Hasbi Rabbi," Sami Yusuf	New Year's Day at home

	Where participant's song story recorded	Participant's nationality	Musician's nationality	Chosen track/title and artist	Song story focus
8	Greece	Lebanon	Lebanese	"Mwaffaq Ya Aaskar Lubnan," Fairuz	Lebanon in the snow
9	Greece	Syria	Syrian	"Ya Hef," Samih Choukeir	The war in Syria
10	Greece	Egypt	Lebanese	"Alli Gara," Wael Jassar	Alienation/far from family
11	Greece	Ghana	Ghanaian/British	"Woyaya," Osibisa	Not knowing where you are going
12	Italy	Romania	Romanian	"Galbena Gutuie," Florian Pop	Grandparents' quince tree
13	Palermo, Italy	Gabon	French	"Lune de Miel," Don Choa	Overcoming difficult times
14	Italy	Lithuania	Azerbaijan	"Nocturne," Muslim Magomaev	Grand parents
15	Italy	Ukraine	Ukrainian	"Ca va," Talita Kum	"We are all foreigners"
16	Italy	Tunisia	Tunisian	"Mahboubi," Abdelwahab El Hannachi	Grandmother
17	Oslo, Norway	Iran	Italian	"Melodramma," Andrea Bocelli	Self
18	Norway	Bangladesh	Pakistani, Bangladeshi	Sufi music (e.g., Nusrat Fateh Ali Khan)	Self
19	Norway	Germany	Norwegian	"This Town," Kygo (feat. Sasha Sloan)	Belonging
20	Norway	Senegal	Senegalese	Pannekake (own song)	Positive message
21	Spain	Morocco	Spanish/Moroccan	"La tarara—Bent Bladi," Orchestra Chekara	
22	Spain	Hungary	Hungarian	". . .," Ghymes	Self
23	Spain	Austria	Austrian	". . .," Arik Brauer	Memories of childhood
24	Spain	Italy	British	"Spread Your Wings," Queen	Hope
25	Spain	Germany	Argentinian	"No soy de aqui, ni soy de alla," Facundo Cabra	Family, belonging
26	Spain	Germany	German	"Tsen Brider," Zupfgeigenhansel	Makes me happy

(*Continued*)

Where participant's song story recorded	Participant's nationality	Musician's nationality	Chosen track/title and artist	Song story focus
27 UK	Iran	Iranian	1. "Gole Royaayee," Omid Soltani 2. "Sornaye Nowruz," Rastak Group 3. "Tasavor Kon," Siavash Ghomaysh	Family, fun
28 UK	Morocco	Welsh	"Holding Out For a Hero," Bonnie Tyler	My driving test
29 UK	Lithuania	North American	"Don't Stop Believing," Journey	Home

It should come as no surprise that reminiscence was core to many stories that emerged, in keeping with research on the topic (Istvandity 2019; Gardner 2022; Cohen, Grenier, and Jennings 2022) and my own experience of non-music generated digital storytelling. The urge to make sense of a past, now at geographical as well as temporal remove, was strong across the stories. Some of those moves were precipitated by war, some were shaped by work or education, but LaBelle's point seems to be relevant to them. He wrote that "To take flight is fundamentally an act of psychic labor, punctuated with dreaming and loss" (2018: 96) and some kinds of "dream worlds" came to the fore in these stories: fragments of the past, slices of time now spliced with the music that either precipitated them or were wrapped up in them. As we knew from digital storytelling, many narrated how the song they chose "reminds" them of a place or a person/people (Istvandity 2016), saying much about the loss that was still felt in the present (Auge 2008: 9), while three more broadly used De Nora's (2000) idea of music as a technology of the self to narrate their relationship with songs.

Ten Song Stories

Ten Song Stories are presented here. I realize that by choosing some over others, I am acting as curator, with all that entails in terms of choice and erasure. These choices are subjective. Some speak to me more than others because the stories affect me, or the music has a "felt materiality" (Garcia 2015) that moves me; makes me dance. Of course, without the music, without the act of the double listening that is possible through the app, all you have is the words. So, as you read, open the MaMuMi app and listen to these ten stories on the app at www.mamumi.eu. This is an effort to counteract the "migrant silencing in European media" (2022: 151) that Chouliaraki and Georgiou report, these stories are presented without too much from myself and to avoid the process of speaking "for," a "process, [where] migrants are silenced or ventriloquized, and it is western 'experts' instead, whether politicians or support organizations, who become the voice trusted to speak on their behalf" (2022: 151). There is no "on behalf" here, all we hoped to be doing was listening and creatively engaging (Western 2020). I offer some context in terms of the song, its singer, date, its lyrics, and its "feel" in order to foreground some of these "erased and inaudible [migrant] voices" Voegelin (2019: 114). I have aimed for a selection in terms of location: of where the recordings took place, Sofia, Bulgaria, Nicosia, Cyprus (3), Triana, Spain, Palermo, Sicily (2), Greece (2), UK and refer to others in passing. I have not edited the responses; these are the same as the transcriptions on the app. They are presented without much theoretical critique as I want the experience to be like moving across a dial and hearing different voices sound out. They are grouped loosely in broad themes, Songs 4, 24, and 12 are about loss and the past, songs 16 and 8 are about family, song 28 is where music acted as a "technology of the self," songs 1, 4, and 15 are about journeys, or being a "foreigner" and songs 6 and 9 are included for the discomfort that connected listening generated.

Song Story 4, "Tamala," (1998) by Foday Musa Suso is from a young man who migrated from Gambia and was recorded in Nicosia. It is a very clear example of the use of music acting as a "time collapser," bringing the past into the present via the memory act and enunciation. Running at 11.51 minutes, the "call and response" based song is about traveling and xenophobia. The participant says how:

> This piece of music is important to me as a memory because it reminds me back home in the Gambia when myself and my friends we used to play "djembeh" while singing this piece of vibes, this piece of music on the mango trees or on the Mahogany trees, in the streets, in the ghettos, so it's a very good reminder.

This short clip from song 24, "Spread Your Wings" (1977) by Queen, a story recorded in Triana, Andalucía, is revealing another form of loss, this time of youth. The 1977 ballad by Queen starts with a piano solo and builds to the operatic rock for which they are famous, Freddie Mercury's vocals singing about "spreading your wings." There is something wistful here, the knowledge that something has gone forever, but again, which the song can usher into the present.

> A song I particularly like is a song by Queen called "Spread your Wings" which is from the late 1970s, early 80s. Because it brings me back to the time when I was younger and I lived in London, and everything was absolutely carefree. And I had the opportunity to do whatever I wanted and meet people.

Song Story 12, "Galbene Gutuie," a traditional Romanian song covered by Florian Pop (2020) was recorded in Palermo. It is about the narrator's childhood, evoked by the memory of the grandparent-child relationship which here circles around a garden and in particular, a quince tree. The lyrics are by Romanian poet Paunescu (1943–2010) and use the quince as a metaphor for home and change.

> Yellow quince . . .
> Bitter-sweet . . .
> Lamp at the window
> All of our winter!
>
> Sweet, yellow light
> Like blond I was.
> Mother put a quince
> That is slowly ripening by the window . . .
> I would bite it, but I'm hurting,
> A kind of yearning takes over me . . .
> And now when the year passes
> It's as if I feel it aging
> . . .
> My mother had no globes
> And no tinsel and no star . . .
> With quinces
> The winter holidays she decorated . . .
> Mother put a quince
> At the window to the road . . .
> And I see how it illuminates, I don't have the power to consume it.[10]

---

[10] https://lyricstranslate.com

The story that weaves in and out of this musical poem is cross-temporal and multi-sensorial; it has taste, it has smell, it almost has a heat to it; the garden in summertime, and then the mellowing as the quinces fall in autumn:

> There is a song which has a particular meaning from me and that I remember with pleasure. The title of this song is "Galbene gutuie," the composer is Adrian Paunescu. It is a song that reminds me my childhood. Galbene gutuie in Italian means "yellow quince," it is a strange fruit, I don't know, many people don't like it, but it is a fruit which has a particular flavour. The song talks about the emotions and the feelings that I felt when I was used to go to my grandparents' house. They had a quince tree in front of their house and we picked up these fruits in Autumn when they dropped. As they have a long ripening time, we picked them and we were used to put them in the house on the window sill and we did not want to eat them. We liked to stare at them for their colour and the smell they had, so we did not want to eat them. We ate them when they were very ripe. And this song talked just about this, these emotions and feelings. So, this song brings me back to my past. I gladly remember these past moments. I think about my grandparents' house, I spent so much time with them. So, when I buy or I see a quince in the local market or supermarket I recall those memories. They are vivid and real memories. I connect this song with my childhood, adolescence, youth.

Song Story 16, "Mahboubi," sung by Abdelwahab Hanachi was recorded in Palermo over Skype and is the only one to refer to the pandemic. It is a popular Tunisian song, sung in Arabic. The narrator was not able to spend the holy month of Ramadan with her family and she describes her frustration at this, before moving on to describing the togetherness of family occasions.

> I listen to a little of everything. It depends from the periods in my life. Often, I listen to Tunisian folklore because I am home sick. But I don't listen to a specific kind of music. I like Arab music world, Palestinian music (which for me it is the number one, I have two favourite singers) which I think is the best, not so much Italian music to be honest.

> It is a song from the Tunisian repertoire, very elegant, beautiful, I liked the new arrangements done recently because it is an old traditional and classic song. "Mahboubi" means "my beloved." In this text the author describes his love with the objects in his house because the beloved one is far away so she looks like that tree, those little branches, those leaves. The words are very elegant and it's a week I am obsessed with this song.

> I feel that I am stuck here in Palermo as recently from April I rented an apartment here due to the pandemic as I am working here and I cannot go back home,

especially during the Ramadan which for me it's something new, so music and folklore makes me feel better and calm. So, I don't think I am alone. My grandma was used to sing in weddings these songs from the Tunisian tradition, oldies. My grandma is an amazing singer. She has an amazing voice and brilliant sound. So, in weddings we are used to sing: we don't need a band, we sing. In these nights there is no need for karaoke. We bring percussions, other instruments and we sing all together. We have always been singing in our house in Tunisia.

Song Story 8, "Wakef Ya Askar Lubnan," (1979) is by renowned Lebanese singer Fairuz (b. 1934). The singer has been recording since the early 1950s and has icon status across the Middle East. The narrator here uses it to talk about the links between herself and the family that now live far from here in Lebanon:

> I am about 21 years here (in Crete). But, whole heart, as a whole, is in Lebanon, in my brothers, in my village, in my friends, in my relatives. I am in constant contact with all my friends, God bless Viber, WhatsApp and Facebook that we can be in daily contact with. Facebook in order to see how they grow up, like my nieces, nephews, their children.

Song Story 28, "Holding Out for a Hero," (1984) is by Bonnie Tyler. I recorded this over Zoom and had known the participant through previous research projects. They had grown up in Morocco and were now working in the police force. The story is about how they remember listening to the song before their driving test, which they took in Melton Mowbray, in the English Midlands.

> It was my driving test and I have the album. I was working with my brother-in-law in the shop while I was doing university and college and stuff. So, the time when I my driving test first time and I failed basically because I was careless, you know, because I was, I had my own car before I had a driving license, you know? I was driving around . . . I learned all the bad habits, but I just needed some kind of, uh, aspiration. I needed some bit of a positive attitude and, you know, and just to chill. And I played that song and I liked it that time.

Some participants told stories explicitly about the journeys they had undertaken and in voicing those stories they became the owner of that journey. The next narrator might be called a "sonic figure[s]," "the itinerant is a product of its surroundings and travels; and what it carries forward is an assemblage of interactions" (LaBelle 2018: 109). The sonic figure here is a young Breton living in Bulgaria, whose Song Story 1 is unique in that he sings it himself, switching between talking and singing. Saying how the song, though old, is relevant today to people who leave their homes.

> Hello, my name is M.... I come from France and I'd like to share a song called "An Hini A Garan." "An Hini A Garan" is a traditional song from Bretagne, a region of France. And more specifically it's called a "gwerz." "Gwerz" is a type of written original music, a kind of poetry sung in the Breton language about melancholic feelings—like a lament. And it is always written in a very personal and intimate way, so that the audience can always relate to the emotions that are expressed in the songs. "An Hini A Garan" means "The One I Love," and it is the lament of a person expressing their sadness about how a long-time beloved one has left for faraway lands to find work and will never come back. It is a song that is important to me because I have myself returned to the roots and it feels to me as an expression of my cultural legacy. And even though I've been living most of my life in Parisian region, it connects me to this part of my roots, which is something that is important to me. It's a song that speaks to me, even though I've never lived such a similar story. However, it is like the story of my grandparents who left their home region of Bretagne and their family to find work in Paris. But it's also the story of many people across the world who have left everything behind them in the hope to find a better life, and the story of all those who stayed behind.
>
> So, it is a story that many people can relate to, and even without understanding the language, you can still feel the sadness carried by each word of this song. When I listen to this song, it doesn't remind me of any specific event or memory of my life. However, whenever I've been living abroad on my own, in a strange way, listening to this song brings back to my mind the image of Breton shores, the feeling of the sea and everything. He made of, um, Britain shores, the feeling of the sea and everything that goes with it. So in a weird way, it makes me feel homesick. But not for the region where I'm, I've been living or something my life. So it is very personal to me and this is why it's also so song that I like to sing whenever I feel sad as a way to express my emotions. So I hope that it is a song that you will come to enjoy.

Song Story 4, "Lampedusa," by Sidiki and Toumani Diabate, is from a young Gambian musician whose story was recorded in Cyprus. He talks about the significance of the song to him:

> Hello everyone, how are you doing? I am happy to be at the MaMuMI project and I am a young migrant living in Cyprus—almost 3 years now I am in Cyprus. I am a musician and percussionist, I play drums, I like instruments that produce

melody especially the KORA—is an African 21 string instrument. I like the song from Toumani Diabate and his song called Lampedusa. It is a song composed about Lampedusa, an island in the Mediterranean, close to Italy and they made this song because a lot of migrants travelled from Africa to go to Italy and they couldn't make it, lot of people died reaching Lampedusa like this. This song reminds me of my journey from the Mediterranean travelling to Cyprus, the difficulties and the sorrows and the song reminds me also of my mother and the song reminds me of a lot of things about Africa, about the migrants, about the youth that are trying to cross either the Mediterranean or the Atlantic Ocean that go to secure their lives, that go to a different world, to a different culture to meet different people, to get a new beginning of their life.

Song Story 15, "Ca va," by Talita Kum (2002) is a vocal-driven pop song, with guitar and oboe. Kum was the Eurovision entry for Ukraine in 2005, coming in tenth place. The participant says they chose it because the lyrics mirrored her own experience:

My name is M.... and I am from Kiev, Ukraine and I am here in Palermo for about 7 years. At first, I was a volunteer with the European Voluntary Service. Then I made my PHD here and now I am staying here because I work here. There is one song by a Ukrainian band that does not exist anymore, I think. It is a song from 2002. I remember very strongly why I think about this song: when I first came home from Italy, my first time here, I was staying home for six months and I was applying for a PHD and I didn't know if I am going to stay in Ukraine or I am going to come back to Palermo. For me this song was a little bit painful but also nice to listen because it's called "foreigners" and more or less the song, the lyrics say "we are all foreigners" somewhere. Everywhere we are foreigners, it depends on our life stories, whether we live in our country, whether we don't live in our country. You can live in your country but in another city so you don't feel you are from there. For me it was very relatable at that point, especially because after some time abroad you come back to your country and you realise you don't fit there anymore also but in the other country you also don't fit because you are a foreigner.

There were two stories that were discomforting, in that they referred to traumatic events. Our Cypriot team lead recorded Song Story 6 with a woman whom she had worked with for a number of years. The woman had moved to Cyprus from the Netherlands and chose the 1978 disco track "I Will Survive" by Gloria Gaynor. Her narrative centers on her emotional responses to what seem like problematic heterosexual relationships, and how she managed to cope with loss.

The song "I will survive," I choose, and it's a song which is in my life regularly popping up again and I love dancing. I get a lot of energy with the text "I will survive," especially after I lost a man, it gives me a lot of power. For me it is not easy to lose something. I had a very bad accident and I lost everything, and then I could not cope with it and then I became easily depressed. I lose a man who was close to me or a child of a man, then it is not easy for me to really go on with life. But this song "I will survive" always gives me a lot of strength—in the past I could not get that strength and I could not survive. So, the song gives me a push and the last 2-3 years it is for me easier to cope in a healthy way with loss in general and this song is for me—how do you say?—a key for that.

This encounter prompted our Cypriot team member Annita Tsolaki to reflect on the process, and in her email to me on July 5 she set out what had happened:

I am sending you this email to share my experience with the mamumi workshop. Of course, this was not something I could anticipate and it is a very sensitive subject to deal with without the necessary expertise. If there was a psychologist with us perhaps, this could have been treated in a rather different way. What I could do is to hear her out and believe her, making her feel safe in sharing this.

Therefore the mamumi workshop worked as an opportunity for people to share their feelings and feel safe while at the same time it had the limitation of how to react when you hear certain sensitive things that require professional treatment in order not to escalate the trauma and damage the mental health of the person participating.

There are often unintended outcomes in applied research, and this proved to us that more training would be needed at the outset, and were we to run the model again, we would add into our manual, how there might be unforeseen emotional encounters. The following story exemplifies this. It is a story of a young woman who had recently arrived in Greece from Syria. She chose a song "Ya Hef" by a Syrian musician Samih Choukeir, who is now in exile in Paris. The lyrics are a direct criticism of the brutality being waged across the country, and this brutality is mirrored in the short narrative this participant provided to the member of the Greek team in Athens. It is a visceral echo of the news reportage which was coming out of Syria, especially in the Battle for Aleppo 2012–16. I was going to open a co-authored article with Kai Arne Hansen with this, but then, on his prompting, wondered if this was a "voice[s] filtered through an aura of victimhood" (Western 2020: 304) and replicating, rather than echoing, the dominant news discourses around the Syrian war. After some reflection, I have put it in here because it is the only story that uses a politically motivated protest/poetic song to talk about the violence they have witnessed. It is part of the MaMuMi Song Story collection. It is discomforting to hear and because of that, I offer it up here. Choukeir's team in Paris sent me a translation of the song. The first verse goes like this:

> Alas! Oh, alas.
> The deluge of bullets on defenseless people. Alas.
> How do you lock up young people at the tender age of roses?
> You, child of my country, murderer of my children,
> You turn your back on the enemy and threaten me with your sword!

The story around this song, recorded in Greece, is very short. Its curtness masks the brutality it narrates. This is what is being carried in this one person's voice, a trauma which was being vocalized. I find it very hard to listen to, its brevity belying the intensity of awfulness and loss that it reports.

> I have sad and bad memories from this song because when the youth of our area went out to protest, they thought that the policemen would not attack them, but they hung them and they killed them. I remember that half of my family died in this situation. After that, we tried to escape from this area and in the middle of the way my mother was stabbed; she was in a bad situation. Because of that, every time I listen to this song, I have bad memories.

Trauma and violence reverberate through this story and heard together with the song makes for a discomforting experience. Sandwiched in between stories about missing families in Lebanon and Egypt it fractures any sense of ease. It is worth remembering too that "Many people on the move are coming from a situation where 'speaking out' or 'raising voice' brought with it the risk of persecution, disappearance, death. Voices carry the border within their timbres. Voicing presence in displacement is courageous as well as creative" (Western 2020: 304). Voice mapping is now another surveillance technique, being used in tandem with biometric data, iris scans, and so on (Bradley and De Noronha 2022: 131). This voice, with this song, is a moment of bravery and testimony that now sits in this small collection of migrant voices. This one voice in particular brings the viewer/listener on the MaMuMi app or website closer to that violence that has propelled them into migrancy. Theirs is the voice of the presumed "necro-political subject," here, the wandering non-person, the "emblem of the contemporary necro-power, because they are the perfect instantiation of the disposable humanity" (Braidotti 2013: 127). This voice, this story, comes out of a refugee camp in central Athens, what Diken calls "sterilized, monofunctional enclosures" (2004: 91), part of the "inhuman face of Fortress Europe... undignified monuments of posthuman inhumanity" (Braidotti 2013: 127). From this necro-place comes the short story supported by the singer, who too has fled and is in exile. This song story, like the others, is an "accretive (life-building) gesture" (Berlant 2007: 757), as it adds voice to the silent journeys and places of trauma, which in this case was precipitated by the flight to Greece. By claiming a small sonic space, the telling of and listening to provides a gesture of exchange and witnessing that disrupts the silenced figure of the "migrant" and adds a resonance to it through voice and accompanying music.

These Song Stories add sound where there is silence, offer individual experience and story to numbers, figures and what we hoped would happen, was that they could, through this specificity and ordinariness, counteract prevailing discourses that render the migrant homogenous and either a threat or in need of saving (Braidotti 2013; Chouliaraki and Georgiou 2022). If voice is resistance (pp. 149, 153) in order to counteract the "constitutive voicelessness" that is Spivak's (2010) reading of colonial subjectivity, then these stories resist the muteness that dogs the contemporary migrant, who is often spoken about and for, but is largely silenced. The person on the move, the unwelcome outsider is very often the owner of an "erased and inaudible voice[s]" (Voegelin 2019: 114) where "silencing and generalizing are part of a western economy of voice" (2022: 123) to which, like the economy itself, the migrant has no access. In his work on sound and displaced listening in Athens, Tom Western asks us to be mindful that "refugee voices" stories serve to abstract and homogenize by depoliticizing displacement and rendering refugees as an ahistorical human category (2020: 304). If anything, the stories here resist depersonalization and dehumanization; they are narratives of affinity, telling recognizable and/or shared experiences of family, loss, and belonging.

In foregrounding these affinities, MaMuMi has allowed for some disruptive work to be done, some "shifting" in its emphasis on listening to voices whose silence is generally secured beneath the noun and epithet, "migrant." Chouliaraki and Georgiou (2022: 108) note how the migrant "alternates or co-exists" (109) as both threatening, with agency and malice and also, as vulnerable, the victim. The stories we heard moved beyond those categories that only make sense in opposition to the "placed," "bordered," and legal, citizens of boundedness. They had recognizable trajectories, of individual journeys and of affective ties. The stories hint at fuller lives than is housed within the word "migrant"; childhoods spent in cherished places, with grandparents; mothers getting drunk around tables, gardens that were grown up in, grandmothers who sang, problems that were overcome. These are shared stories enabled through the singular voice that delivers the story in combination with the interwoven fragments of song. We made the decision to subtitle the stories in the partner languages rather than dub them, as it was important to hear the voice. One of the reasons for doing this was that we wanted ourselves, the academics, and NGOs to be voiceless for once, to be mute. It was important to not dub the voices on the app after considering what Chouliaraki and Georgiou have to say about the types of voices that might emerge when working with migrants:

> Ventriloquized voice refers to contexts where migrant voice was appropriated within western discourse, echoing, mirroring, and legitimizing its securitizing agendas, and articulated voice is about contexts that attach value to migrant voice enable it to be heard as an equal in public conversations. While articulated voice fosters forms of social and political recognition, we have shown here that ventriloquized voice offers visibility but no recognition. (Chouliaraki and Georgiou 2022: 69)

The MaMuMi participants articulated small slices of their own lives, by narrating their stories in English, German, Arabic, Farsi, Greek, and Italian: their first or second language. These voices come through the app, witness to movement and yearning, voicing the in-between places.

Impressions

> How do we understand the voice then, this voice out of place and without legal status? How do itinerant voices resound within the contemporary Western environment? Explicitly those marked by African or Arabic origin, and yet grounded within European territories today? Meeting others, giving testimony to their journey, and prophesizing about a future without borders—what are the lessons brought forward when encountering such vocalizations? (LaBelle 2018: 112)

LaBelle asks the question that we on the project team also asked, although many of our participants had legal status, having migrated to their current homes many years previously. MaMuMi was an attempt at adding volume to those who are silenced and at individualizing those who have been depersonalized. The listening happened at the workshops, at the moment of interview/recording and then again via the app. It had, what Bickford calls "a way of breathing life and story into a person who has been categorized. It is about inclusion" (Bickford 1996). It set out to be such a "thing," that is, an inclusive tool for exchange and awareness. And as such, it added another strand to the participating NGO's suite of approaches to working with migrants and in doing so had differing degrees of impact. In Cyprus, the NGO reported that it increased awareness of the possibilities for intercultural exchange and integration provided by conversations about music, and local stakeholders were enthusiastic about pursuing new opportunities for adapting and implementing the MaMuMi song workshop methodology. What was especially useful about the approach was that it extended the supply of high-quality learning opportunities tailored to the needs of individuals they are working with. The Italian partners reported that it was the platform (the app) that was an important factor in the possibility of enabling minority groups to share and spread their migration experience to other communities, groups, schools, and civic organizations. They also thought that the MaMuMi Migration Map could achieve a wider impact given its potential to be used at the origins of migration journeys and NGOs working at these geographic points. In Norway, the Agency of Cultural Affairs in Oslo expressed interest in implementing the MaMuMi workshops in local meeting places for migrants, such as libraries or "cultural cafés." And our final report documented how our dissemination strategy had exposed around 150,000 people to the project across the NGO sector, cultural organizations, and academics.

There were unforeseen repercussions the workshops had produced that required further ethical considerations, especially for those conducting the group workshops and individual recordings. They were to do with the potential for what LaBelle calls "listening . . . being conducive to empathy and compassion" (2018: 4). The Cypriot partner reported back to us that some of their participants had used the workshop to share other life experiences such as violence or abuse. This exemplifies Stewart's ideas on "ordinary affects" whereby "it shows where things can go, taking off in their own little worlds, when something throws itself together" (2007: 40), this "something" being the collision point where the unlikeliness of an NGO asking a recent or long-term migrant about a song they like has the ability to "take off." This has been positive; responses from the NGO teams in Cyprus and Greece especially noted that there had been some impact on the NGOs themselves, as those working with migrants told them how therapeutic the MaMuMi workshop was. Annita Tsolaki, the project lead in Nicosia, noted in an email on July 21, 2022, in response to my query on how the model was working for them

> What I can say about the mamumi app is that a lot of people have been telling me how successful and impactful our project has been and that they are using it as a best practice example. We plan to use them in the future perhaps at another workshop/event we will be doing related to the topic of music, storytelling and integration. (Tsolaki July 21, 2022)

Some of our initial objectives were perhaps greater than could be achieved in a small two-year project. We had modest goals based on the nature of our professional networks and organizations. We were not working with policy-makers but were trying to engineer spaces where conversations could be had that broadened out the life-worlds that NGOs might encounter, through the use of listening to others talking about music. This I think we did in a small way. And I am grateful to have been offered the chance to work across those borders because as Britain voted to leave the European Union in 2016, the UK cannot lead any more Erasmus+ projects, cannot collaborate with its European partners, cannot share. This move is coloring my reflection and review of the project, as I see the borders of my own country hardening up and closing in.

Conclusion

The MaMuMi project created affordances between listening, story, and song. It worked on the premise that we all have "narratable selves" (Cavarero 1997, in Chouliaraki and Georgiou 2002: 110) and the narrating of this in a public realm literally constitutes a new voice to be heard, a presence to be counted, a sonic slice of humanity speaking out. And this last word is so very important to the design, motive, and implementation of the MaMuMi Song Story project, given that we wanted to share stories about humanity that bled across borders of categorization and naming. This is not to erase difference. It does not eradicate understandings of material disparities in life experiences and conditions. But it does argue for spaces of exchange, facilitated through the connected listening that storytelling about song enables. This enablement comes from the collapse in distance that the listening moment affords, it is again a matter of proximity. We come into close contact via the voice and the song to a moment revealed, a person remembered. The listening event is about "nearness" (Auge 2008: 7) and extends the possibility of convivial listening through the song story, a "sounded act."

> The limits of bodies and things are radically extended through sounded actions, making of them expressive flows open to intersections and overlaps as well as fragmentations and ruptures. (LaBelle 2018: 61)

I am not sure our work was radical, but it was useful. It was an end in itself, "performing voice-as-resistance in broader networks that intermediate migrant experience in the global North" (Chouliaraki and Georgiou 2022: 153). We came close to people through the listening; we affected affinities. We punctured the skin of the word "migrant" and listened beyond and within it. I like to think that it ruptured the silence behind which migrants are forced, enabling a more "personal perspective" to be enfolded in, and so we colored in the lives behind the word "migrant" with stories about songs, and triggered by songs. The listening encounters that we enabled went some way toward affecting shifts and extending connections with people. It opened up routes to inclusivity through shared listening and the exchange of stories about songs. We made new zones, where we listened together to a song, and listened to someone talk about that song. The stories were about loss, family, memory, and movement. Between us, we carved out novel lines of identification and affective connections. LaBelle talks about the "rustlings and stirrings by which listening and being heard takes place" (2018: 61). Like Stewart (2007), these shifts might upset, or "rupture," doing so in a way that problematizes the fluidity with which migrants have been framed, giving them instead the fixity of ordinariness, family, loss, and humor, vivifying them through narrative and song. For the seven people who were on the plane that did not go to Rwanda, there was no such space to be heard.

6

Echoes

This has been a book about listening to voices. It has revolved around silence and ordinariness. It has brought listening into dialogue with memory and belonging into relationship with listening. It has focused on the small, micro-engagements that crouch within the superstructures of violent border control and the censorious policing of sonic citizenry. It is on the side of the sonic agents, the reshuffling of histories and hierarchies afforded by acts of listening, of ruptures through resonance.

It has used story as the crutch upon which to tell and to listen: big stories reframed on social media in small vignettes, and personal stories retold in performative historiographies. It has listened with feminist care, that is, it was my body that was brought into the listening exchanges and this body came with a feminist sensibility that prioritizes care as giving space to talk and be listened to. Turn the sound up, let's hear those who have been silenced, and let them be listened to. It is enough to listen. This might make all the difference. Micro within the macro.

We are moored through sound. Listening does not happen in a void. Mapped into our bodies, our world, our time, the listening we do comes from those embodied places that lie at the complex intersection/imbrication of lived experience and cultural context. Listening is relational, it puts us in place, cast into a relationship with others, ourselves, our world through being (and not being) listened to. At the same time, listening rips through time, prizes us out of our bodies into affective intimacies with those whose stories we don't know, have never listened to before, who call us from the past. Listening brings us to the edge of meaning and tips us over the edges of time as it too has been mapped into a neat linear line.

The listener is imbricated with but different from that of an audience or a fan: a lonely figure within Western culture; a spy, an eavesdropper, a confidante. A docile body, the good listener. What does it mean now, to "listen"? What is it about these projects, and this move to listening that I have considered important enough to write about? Georgina Born asked similar questions in 2010:

> we should begin by asking, what is it to listen, and how should we conceive of listening? Or: how should we frame, or what are the boundaries of this activity that we call listening? Does it take place within the mind, or (also) within the body? Is it something that is primarily individual in its operation, or that is socialized and encultured? (Born 2010: 80)

In answering these questions, Born focuses on musical experience. I focused on talking, noise, and silence. I worked with "veterans" but listened to Alan, Rick, Sally, Willy, and Chris. I saw "victims" who were murdered in the Holocaust on my Twitter feed but listened to the stories of Suzanne, Genevieve, and Leon. I heard the daily discussions over "illegal migrants" and asylum seekers in the UK, but listened to the stories from Maryna, Daviny, Jelena, and twenty-six others whose names were known to their interviewers/listeners. All of us on these research projects, myself, and the teams in the UK and across Europe, have listened across age, time, and borders. This listening has been shaped by story, and these narratives fill out the world and make it sound fuller.

Turn It On. Listen In

Let's listen to one more story from Mapping the Music of Migration because it conjures up voices on the edge of that project (Figure 6.1).

Song Story 11
From Ghana to Greece
"Woyaya" by Osibisa (1971)

> The song talks about life and its uncertainties, not knowing where you are going. Today you are very happy, like me, yesterday I was in my home country, I felt safe, I was in my bed and all of the sudden you have to go somewhere, you don't know where to go. This song tells you despite all this that there are other ways, there is hope. Back home when I was teaching in Grade 4, we had within the curriculum to learn this song. The background of this song, the meaning and everything, the children understood very well, so every time I listen to it, I remember of them.

This is a photograph of a travel document: a steamer (ferry) ticket for a "Boy," dated December 27, 1938. The boat is named (Prague) but the 'boy' is not. This is

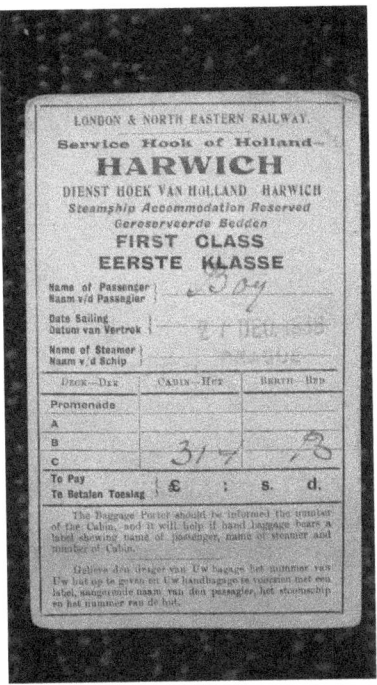

Figure 6.1 Robert Landberger's boat ticket from Holland to England, 1938. © Synagogue Lane 2009.

a photograph of a ticket kept by Robert Landberger, an Austrian Jewish boy who, at the age of twelve, had been put on a Kindertransport train going from Vienna to the Hook of Holland, where he would travel on to Harwich in the UK. Robert was a member of the Cheltenham Hebrew Congregation and was taking part in a research-led film about the members of that community who had been on the Kindertransport. Robert did not want to be filmed, so myself and the team used digital storytelling methods with him, recording his voice as he narrated the tale of his leaving Austria. Along with his ticket, he had kept a placard that had been hung around his neck with the number 158 on it, a faded suitcase label with his name on it, and a checklist of the items of clothing he had with him in a small suitcase, which his mother had packed. The four objects that Robert used as memory props to tell his story were mnemonic and have an anonymity to them that is mediated by Robert's voiceover. Without his audio, he would be absent and only his gender, number, and clothes would be testament to him. If you have time, please take that time, and listen to him.[1]

---

[1] https://www.cheltenhamsynagogue.org.uk/

Bibliography

Adorno, Theodor. 2002. "On the Fetish Character in Music and the Regression of Listening." In *Essays on Music*, edited by J. M. Bernstein, 288–317. Berkeley: University of California Press.

Ahmed, Sara. 2004. "Affective Economies." *Social Text* 22 (2): 117–39.

Ahmed, Wasim. 2019a. "Using Twitter as a Data Source: An Overview of Social Media Research tools." Academic. June 18, 2019. https://blogs.lse.ac.uk/impactofsocialsciences/2019/06/18/using-twitter-as-a-data-source-an-overview-of-social-media-research-tools-2019.

Altmay, Ayse, María José Contreras, Marianne Hirsch, Jean Howard, Banu Karaca, and Alisa Solomon. 2019. *Women Mobilizing Memory*. New York: Columbia University Press.

Apfelbaum, Erika. 2010. "Halbwachs and the Social Properties of Memory." In *Memory: Histories, Theories, Debates*, edited by Susannah Radstone and Bill Schwarz, 77–92. New York: Fordham University Press.

Appadurai, Arjun. 1986. *The Social Life of Things: Commodities in Cultural Perspective*. Cambridge: Cambridge University Press.

Auge, Marc. 2008. *Non-Places: An Introduction to Supermodernity*. London: Verso.

Avelar, Idelber, and Christopher Dunn. 2011. *Brazilian Popular Music and Citizenship*. Durham: Duke University Press.

Back, Les. 2007. *The Art of Listening*. London, New York: Bloomsbury.

Barad, Karen. 2010. "Quantum Entanglements and Hauntological Relations of Inheritance: Dis/Continuities, SpaceTime Enfoldings, and Justice-to-Come." *Derrida Today* 3 (2): 240–68.

Barad, Karen. 2017. "Troubling Time/s and ecologies of nothingness: re-turning, re-membering and facing the incalculable." *New Formations* 2017 (92): 56–86.

Barbour, Charles. 2011. *Philosophy and Social Criticism* 37 (6): 629–45. https://doi.org/10.1177/0191453711402938.

Barnd, Natchee Blu. 2017. *Native Space: Geographic Strategies to Unsettle Settler Colonialism*. Corvallis: Oregon State University Press.

Barthes, Roland. 1977. *Image, Music, Text*. Hammersmith, London: Harper Collins.

Barthes, Roland. 1985a. "Listening." In *The Responsibility of Forms*, 245–50. Berkeley, Los Angeles: University of California Press.

Barthes, Roland. 1985b. *The Responsibility of Forms: Music, Art and Representation*. Berkeley, Los Angeles: University of California Press.

Bartmanski, Dominik, and Ian Woodward. 2015. *Vinyl: The Analogue in the Digital Age*. London: Bloomsbury.

Barwick, Nick. 2004. "Bearing Witness: Group Analysis as Witness Training In Action." *Group Analysis* 37 (1): 121–36.

Bassel, Leah. 2017. *The Politics of Listening: Possibilities and Challenges for Democratic Life*. London: Palgrave Pivot.

Bassel, Leah. 2022. "A Promise of Listening: Migrant Justice and the London Permanent Peoples' Tribunal." *Race and Class* 63 (4): 35–55.

Bauman, Zygmunt. 1996. "From Pilgrim to Tourist. A Short History of Identity." In *Questions of Cultural Identity*, edited by Stuart Hall and Paul Du Gay, 18–36. London: Sage.

BBC. 2022. "Crime Bill: Lords Defeats for Government's Protest Clamp-down Plans." *BBC News*. https://www.bbc.co.uk/news/uk-politics-60032465.

BBC. n.d. "The Listening Project." https://www.bbc.co.uk/programmes/articles/41rDvmTW0T1JWjXkcvZtMqt/about.

Behr, Eric. 2014. "Something Inaudible: Anthony Burgess's Mozart and the Wolf Gang and Kirsty Gunn's The Big Music as Literary Music through Roland Barthes's Concept of Listening." *MUSe* 1 (1): 114–24.

Bergson, Henri. 2011 [1911]. *Matter and Memory*. Tunbridge Wells: Solis Press.

Berlant, Lauren. 1997. *The Queen of America Goes to Washington City: Essays on Sex and Citizenship*. Durham: Duke University Press.

Berlant, Lauren. 2007. "Slow Death (Sovereignty, Obesity, Lateral Agency)." *Critical Inquiry* 33 (4): 754–80. https://doi.org/10.1086/521568.

Berlant, Lauren. 2008. *The Female Complaint The Unfinished Business of Sentimentality in American Culture*. Durham: Duke University Press.

Berlant, Lauren. 2011. *Cruel Optimism*. Durham: Duke University Press.

Bickford, Susan. 1996. *The Dissonance of Democracy. Listening, Conflict and Citizenship*. Ithaca: Cornell University Press.

Bijsterveld, Karin. 2008. *Mechanical Sound: Technology, Culture, and Public Programs of Noise in the Twentieth Century*. Cambridge, MA: MIT Press.

Bijsterveld, Karin, and Jose Van Dijck, eds. 2009b. *Sound Souvenirs: Audio Technologies, Memory and Cultural Practices*. Amsterdam: Amsterdam University Press.

Birdsall, Carolyn. 2009a. "Earwitnessing: Sound Memories of the Nazi Period." In *Audio Technologies, Memory and Cultural Practices*, edited by Karin Bijsterveld and Jose Van Dijck, 169–81. Amsterdam: Amsterdam University Press.

Bissell, David. 2010. "Passenger Mobilities: Affective Atmospheres and the Sociality of Public Transport." *Environment and Planning D. Society and Space* 28 (2): 270–89.

Bollas, Christopher. 2009. *The Evocative Object World*. London, New York: Routledge.

Borio, Gianmario. 2015. *Musical Listening in the Age of Technological Reproduction*. Abingdon: Ashgate.

Born, Georgina. 2010. "Listening, Mediation, Event: Anthropological and Sociological Perspectives." *Journal of the Royal Musical Association* 135 (Special Issue no. 1): 79–89.

Born, Georgina. 2013. *Music, Sound and Space: Transformations of Public and Private Experience*. Cambridge: Cambridge University Press.

Brabazon, Tara. 2002. *Ladies Who Lunge: Celebrating Difficult Women*. Kensington: University of New South Wales Press.

Bovens, Mark. 1995. *De verplaatsing van der politiek. Een agenda voor democratische vernieuwing*. Amsterdam: Wiardi Beckman Stichting.

Bradley, Gracie Mae, and Luke De Noronha. 2022. *Against Borders: The Case for Abolition*. London: Verso.

Braidotti, Rosi. 2013. *The Posthuman*. Cambridge: Polity.

Brand, Dione. 2006. *Inventory*. London: Penguin.

Brannelly, Tula, and Marian Barnes. 2022. *Researching With Care: Applying Feminist Care Ethics to Research Practice*. Bristol: Bristol University Press.

Brunsdon, Charlotte. 1997 *Screen Tastes: Soap Opera to Satellite Dishes*. London, New York: Routledge.

Bull, Michael. 2004. "Thinking about Sound, Proximity, and Distance in Western Experience: The Case of Odysseus's Walkman." In *Hearing Cultures: Essays on Sound, Listening and Modernity*, edited by Veit Erlmann, 173–91. Oxford: Berg.

Burgess, Jean. 2006. "Hearing Ordinary Voices: Cultural Studies, Vernacular Creativity and Digital Storytelling." *Continuum* 20 (3): 201–14. https://doi.org/10.1080/10304310600641737.

Burns, Lori, and Melisse Lafrance. 2002. *Disruptive Divas: Feminism, Identity and Popular Music*. New York: Routledge.

Burton, Antoinette, ed. 2005. *Archive Stories: Facts, Fictions, and the Writing of History*. Durham: Duke University Press.

Butler, Judith. 1990. *Gender Trouble*. New York, London: Routledge.

Butterwick, Shauna. 2012. "The Politics of Listening." In *Feminist Popular Education in Transnational Debates*. Comparative Feminist Studies Series. New York: Palgrave Macmillan.

Caddick, Nick. 2018. "Life, Embodiment, and (Post-)war Stories: Studying Narrative in Critical Military Studies." *Critical Military Studies* 7 (2): 155–72.

Campbell, Sue. 2008. "The Second Voice." *Memory Studies* 1 (1): 41–8.

Campt, Tina M. 2017. *Listening to Images*. Durham, London: Duke University Press.

Cancian, Francesca, and Stacey Oliker. 2000. *Caring and Gender*. Walnut Creek: AltaMira Press.

Canetti, Elias. 1979. *Earwitness: 50 Characters*. New York: Seabury Press.

Cardinal, Serge. 2002. "Espace Du Son, Espace Du Sens, Espace Du Sentir, Espace Du Soi. Notes Sur Jean-Luc Nancy." *A l'ecoute*, Galilee, 13–45. https://www.creationsonore.ca/wp-content/uploads/2015/10/resumes_serge-cardinal_espace-du-son.pdf.

Carter, Paul. 1992. *The Sound In-between: Voice, Space, Performance*. Kensington and Strawberry Hills: UNSW and Endeavour.

Casey, Sarah, Abigail Gardner, and Philip Rayner. 2014. *Media Studies, The Essential Resource*. 2nd ed. Oxford, New York: Routledge.

Cesaro, Steph. 2014. "(Re)Educating the Senese: Mutimodal Listening, Bodily Learning and the Composition of Sonic Experiences." *College English* 77 (2): 102–23.

Chandler, Rasheeta, Erica Anstey, and Henry Ross. 2015. "Listening to Voices and Visualizing Data in Qualitative Research: Hypermodal Dissemination Possibilities." *SAGE Open*. https://doi.org/:10.1177/2158244015592166.

Charmaz, Kathy. 2014. *Constructing Grounded Theory*. 2nd ed. Los Angeles: Sage.

Chazan, May, and Maddy McNab. 2018. "Doing the Feminist Intergenerational Mic: Methodological Reflections on Digital Storytelling as Process and Praxis." *Forum Qualitative Social Research* 19 (2). https://www.qualitative-research.net/index.php/fqs/article/view/2949.

Chouliaraki, Lilie, and Myria Georgiou. 2022. *The Digital Border*. New York: New York University Press.

Clarke, Eric F. 2005. *Ways of Listening: An Ecological Approach to the Perception of Musical Meaning*. Oxford: Oxford University Press.

Cloonan, Martin, and John Street. 1997. "Politics and Popular Music: From Policing to Packaging." *Parliamentary Affairs* 50 (2): 223–34.

Cockburn, Cynthia. 2004. *The Line: Women, Partition and Gender Order in Cyprus*. London, New York: Zed Books.

Cohen, Sara, Line Grenier, and Ros Jennings. 2022. *Troubling Inheritances: Memory, Music, and Aging*. New York: Bloomsbury Academic.

Collective, Counter Cartographies, Craig Dalton, and Liz Mason-Deese. 2015. "Counter (Mapping) Actions: Mapping as Militant Research." *ACME: An International Journal for Critical Geographies* 11 (3): 439–66.

Couldry, Nick. 2008. "Mediatization or Mediation? Alternative Understandings of the Emergent Space of Digital Storytelling." *New Media and Society* 10 (3): 373–91. https://doi.org/10.1177/1461444808089414.

Courtney, Paul, Leonie Burton, Colin Baker, Isabel Fielden, Abigail S. Gardner, and Fahimeh Malekinezhad. 2020. "GEM Interim Monitoring and Evaluation Summary Report." Project Report. University of Gloucestershire, Cheltenham.

Creed, Barbara. 1993. *The Monstrous-Feminine Film, Feminism, Psychoanalysis*. Oxford, New York: Routledge.

Cusick, Suzanne. 2017. "Music as Torture/Music as Weapon." In *The Auditory Culture Reader*, 379–92. London, New York: Bloomsbury.

Davies, Bronwyn. 2014. *Listening to Children: Being and Becoming*. New York: Routledge.

Davies, Eirlys E., and Abdelali Bentahila. 2006. "Code Switching and the Globalisation of Popular Music: The Case of North African Rai and Rap". *Multilingua - Journal of Cross-Cultural and Interlanguage Communication*. 25 (4): 367–92. https://doi.org/doi:10.1515/MULTI.2006.020.

Davis, Ruth F. 2004. *Ma'luf: Reflections on the Arab Andalusian Music of Tunisia*. Lanham, Toronto, Oxford: The Scarecrow Press.

De Nora, Tia. 2000. *Music In Everyday Life*. Cambridge: University of Cambridge.

Deleuze, Gilles, and Felix Guattari. 2013. *A Thousand Plateaus*. London: Bloomsbury.

Deller, Jeremy [dir]. 2018. Everybody in the Place: An Incomplete History of Britain 1984–1992.

Denning, Michael. 2015. *Noise Uprising: The Audiopolitics of a World Musical Revolution*. London: Verso.

Diabate, Toumani, and Sidiki Diabate. 2014. *Lampedusa*. London: World Circuit.

Diken, Bülent. 2004. "From Refugee Camps to Gated Communities: Biopolitics and the End of the City." *Citizenship Studies* 8 (1): 83–106. https://doi.org/10.1080/1362102042000178373.

Donovan, Kate. 2019. *Listening beyond Radio, Listening beyond History*. Vol. II. Seismograf Sonic Argumentation.

Doughty, Karolina, Michelle Duffy, and Theresa Harada. 2016. "Practices of Emotional and Affective Geographies of Sound." *University of Wollongong Research Online*, Faculty of Social Science Papers.

Dreher, Tanja. 2009. "Listening across Difference: Media and Multiculturalism beyond the Politics of Voice." *Continuum* 23 (4): 445–58. https://doi.org/10.1080/10304310903015712.

Dunford, Mark. 2017. "Understanding Voice, Distribution and Listening in Digital Storytelling." *Participations, Journal of Audience and Reception Studies* 14 (1): 313–29.

Dunford, Mark, and Tricia Jenkins, eds. 2017. *Digital Storytelling: Form and Content*. Palgrave Macmillan.

Dyer, Richard. 1997. *White*. Routledge.

Dyer, Richard. 2013. *White: Essays on Race and Culture*. Routledge.

Edwards, Annabelle, Brigit McWade, Samuel Clark, and Liz Brewster. 2021. "Older Veterans: The Materiality of Reminiscence, Making Unknown Histories Knowable and Forging Social Connections." *Memory Studies* 14 (4): 892–908. https://doi.org/10.1177.

Eidsheim, Nina Sun. 2015. *Sensing Sound: Singing and Listening as Vibrational Practice*. Durham, London: Duke University Press.

Erll, Astrid. 2011. "Travelling Memory." *Parallax* 17 (4): 4–18.

Erlmann, Veit. 2004. *Hearing Cultures: Essays on Sound, Listening and Modernity*. Oxford, New York: Berg.

Erlmann, Veit. 2014. *Reason and Resonance: A History of Modern Aurality*. New York: Zone Books.

Ettorre, Elizabeth. 2016. *Autoethnography as Feminist Method: Sensitizing the Feminist 'I'*. Oxford: Routledge.

Fairuz. 1976. *Wakef Ya Asmar*. Lebanon: Duniaphon.

Fatsis, Lambros. 2019. "Policing the Beats: The Criminalisation of UK Drill and Grime Music by the London Metropolitan Police." *The Sociological Review* 67 (6): 1300–16.

Felman, Shoshana, and Dori Laub. 1992. *Testimony: Crises of Witnessing in Literature, Psychoanalysis, and History*. Routledge.

Fernandes, Sujatha. 2017. *Curated Stories ThGrite Uses and Misuses of Storytelling*. Oxford: Oxford University Press.

Finer, Ella. 2021. "The Commons and the Square: A Politics of Resonance." Edited by & beyond collective. *Sonic Urbanism: The Political Voice*. https://theatrum-mundi.org/library/the-commons-and-the-square/.

Fiske. 2011. *Understanding Popular Culture*. London, New York: Routledge.

Fivush, Robyn. 2008. "Remembering and Reminiscing: How Individual Lives Are Constructed in Family Narratives." *Memory Studies* 1 (1): 49–58.

Forkert, Kirsten. 2021. "'Literature and the Politics of Listening' by Dr. Amber Lascelles." https://bcmcr.org/research/literature-and-the-politics-of-listening-by-dr-amber-lascelles/.

Foucault, Michel. 1975. *Discipline and Punish*. London: Penguin.

Foucault, Michel. 1994. *Aesthetics, Method and Epistemology*. London: Allan Lane.

Fraser, Robert. 2018. *Literature, Music and Cosmopolitanism*. London: Palgrave Macmillan.

Freeman, Mark. 2010. "Telling Stories: Memory and Narrative." In *Memory: Histories, Theories, Debates*, edited by Susannah Radstone and Bill Schwarz, 263–78. Fordham University Press.

Fuente, Eduardo. 2019. "After the Cultural Turn: For a Textural Sociology." *The Sociological Review* 67 (3): 552–67. https://doi.org/10.1177/0038026118825233.

Fuss, Diane. 1989. *Essentially Speaking: Feminism, Nature, and Difference*. New York: Routledge. https://doi.org/10.4324/9780203699294.

Gallagher, Michael, Anja Kanngieser, and Jonathan Prior. 2017. "Listening Geographies: Landscape, Affect and Geotechnologies." *Progress in Human Geography* 41 (5): 618–37.

Gallagher, Michael, Jonathan Prior, Martin Needham, and Rachel Holmes. 2017. "Listening Differently: A Pedagogy for Expanded Listening." *British Educational Research Journal* 43 (6): 1246–65. https://doi.org/10.1002/berj.3306.

Garcia, Luhushis-Manuel. 2015. "Beats, Flesh, and Grain: Sonic Tactility and Affect in Electronic Dance Music." *Sound Studies: An Interdisciplinary Journal* 1 (1): 59–76. https://doi.org/10.1080/20551940.2015.1079072.

Gardner, Abigail. 2015. *PJHarvey and Music Video Performance*. Farnham: Ashgate.

Gardner, Abigail. 2020. *Ageing and Contemporary Female Musicians*. London, New York: Routledge.

Gardner, Abigail. 2022. "Storytelling and Disrupting Borders: A Sicilian Workshop." In *Troubling Inheritances*, edited by Sara Cohen, Line Grenier, and Ros Jennings, 81–102. New York, London: Bloomsbury Academic.

Gardner, Abigail, and Moorey Gerard. 2016, Raiders of the lost archives, *Popular Communication*, 14 (3): 169-77.

Gardner, Abigail, and Ros Jennings. 2020. *Aging and Popular Music in Europe*. New York, London: Routledge.

Gardner, Abigail, and Kai Arne Hansen. 2023. "'Mapping the Music of Migration': Emergent Themes and Challenges." *European Journal of Cultural Studies* Sage.1–14. Doi: 10.1177/13675494231156120

Gaynor, Gloria. 1978. *I Will Survive*. Vinyl. Netherlands: Polydor.

Gorton, Kristyn, and Joanne Garde-Hansen. 2019. *Remembering British Television Audience, Archive and Industry*. London, New York: Bloomsbury.

Gray, Ann. 1997. "Learning From Experience: Cultural Studies and Feminism." In *Cultural Methodologies*, edited by J. McGuigan, 87–106. London: Sage.

Grist, Hannah, and Ros Jennings. 2020. *Carers, Care Homes and the British Media: Time to Care*. London: Palgrave Macmillan.

Grosz, Elizabeth. 1994. *Volatile Bodies: Toward a Corporeal Feminism*. London: Taylor and Francis.

Hagood, Mack. 2019. *Hush: Media and Sonic Self-Control*. Durham: Duke University Press.

Halberstam, Judith. 2005. *In a Queer Time and Place: Transgender Bodies, Subcultural Lives*. New York: New York University Press.

Halbwachs, Maurice. 1992. *On Collective Memory*. Chicago: University of Chicago Press.

Hall, Stuart, Chas Critcher, Tony Jefferson, John Clarke, and Brian Roberts. 1978. *Policing the Crisis Mugging, the State, and Law and Order*. London: Palgrave MacMillan.

Hanachi, Abdelwahab. 2006. *Mahmoubi*. MLP.

Haraway, Donna. 1991. "Situated Knowledges: The Science Question in Feminism and the Privilege of Partial Perspective." In *Simians, Cyborgs and Women: The Reinvention of Nature*, edited by Donna Haraway, 183–201. London: Free Association Books.

Hartley, John. 2000. "Radiocracy: Sound and Citizenship." *International Journal of Cultural Studies* 3 (2): 152–9.

Hashemi, Amin. 2017. "Power and Resistance in Iranian Popular Music." In *Popular Music Studies Today: Proceedings of the International Association for the Study of Popular Music 2017*, edited by Julia Merrill. Wiesbaden: Springer Fachmedien Wiesbaden. https://doi.org/10.1007/978-3-658-17740-9_14.

Havis, Devonya. 2014. "'Now, How Do you Sound?': Considering A Different Philosophical Praxis." *Hypatia* 29 (1): 237–52.

HCA Librarian. 2018. "New to the Library: The Listener Historical Archive." Blog. http://libraryblogs.is.ed.ac.uk/hcalibrarian/2018/07/06/new-to-the-library-the-listener-historical-archive/.

Heath, Shirley Brice. 2000. "What No Bedtime Story Means: Narrative Skills at Home and School." In *Schooling the Symbolic Animal: Social and Cultural Dimensions of Education*, edited by Kathryn M. Borman, 169–89. London: Rowan and Littlefield.

Hebdige, Dick. 1979. *Subculture: The Meaning of Style*. London: Methuen.

Henriques, Julien, and Beatrice Ferrara. 2014. "The Sounding of the Notting Hill Carnival: Music as Space, Place and Territory." In *Black Popular Music in Post-World War 2 Britain*, edited by Jon Stratton and Nabeel Zuberi, 131–52. Ashgate Popular and Folk Music Series. London: Ashgate.

Hesmondhalgh, David. 2008. "Towards a Critical Understanding of Music, Emotion and Self-Identity." *Consumption, Markets and Culture* 11 (4): 329–43.

Hesmondhalgh, David. 2015. *Why Music Matters*. Oxford: Wiley Blackwell.

Hirsch, Marianne. n.d. https://www.postmemory.net/.

Hirsch, Marianne, and Leo Spitzer. 2009. "The Witness in the Archive: Holocaust Studies/Memory Studies." *Memory Studies* 2 (2): 151–70. https://doi.org/10.1177.

Holmes, Oliver, Harriet Sherwood, and Julian Borger. 2021. "Israel Vows Not to Stop Gaza Attacks until There Is 'Complete Quiet.'" *The Guardian*, May 12, 2021. https://www.theguardian.com/world/2021/may/12/israel-only-stop-gaza-attacks-when-complete-quiet.

hooks, bel. 2009. *Belonging, a Culture of Place*. New York: Routledge.

Hristova, Marije, Francisco Ferrandiz, and Johanna Vollmeyer. 2020. "Memory Worlds: Reframing Time and the Past - An Introduction." *Memory Studies* 13 (5): 777–91. https://doi.org/10.1177.

Hrycak, Alexandra, and Maria G. Rewakowicz. 2009. "Feminism, Intellectuals and the Formation of Micro-Publics in Postcommunist Ukraine." *Studies in East European Thought* 61 (4 Wither the Intelligentsia: The End of the Moral Elite in Eastern Europe): 309–33.

Huyssen, Andreas. 2003. *Present Pasts: Urban Palimpsests and the Politics of Memory*. Stanford: Stanford University Press.

Illouz, Eva. 2007. *Cold Intimacies*. Cambridge: Polity.

Ingold, Tim. 2015. *The Life of Lines*. Oxford, New York: Routledge.

Isin, Engin. 2019. "Doing Rights with Things: The Art of Becoming Citizens." In *Performing Citizenship: Bodies, Agencies, Limitations*, edited by Paula Hildebrandt, Kerstin Evert, Sibylle Peters, Mirjam Schaub, Kathrin Wildner, and Gesa Ziemer, 45–56. Cham: Springer International Publishing. https://doi.org/10.1007/978-3-319-97502-3_4.

Islam, Azharul, Elizabeth Sheppard, Martin A. Conway, and Shamsul Haque. 2021. "Autobiographical Memory of War Veterans: A Mixed-Studies Systematic Review." *Memory Studies* 14 (2): 214–39. https://doi.org/10.1177.

Istvandity, Laura. 2016. "'If I Ever Hear It, It Takes Me Straight Back There': Music, Autobiographical Memory, Space and Place." In *A Cultural History of Sound, Memory, and the Senses*, edited by Joy Damousie and Paula Hamilton, 231–44. New York, Oxford: Routledge.

Istvandity, Laura. 2019. "The Lifetime Soundtrack 'On the Move': Music, Autobiographical Memory and Mobilities." *Memory Studies* 15: 170–83.

James, C. 1994. "'Holding' and 'Containing' in the Group and Society." In *The Psyche and the Social World*, edited by D. Brown and L. Zinkin. London: Routledge.

James, Malcolm. 2021. *Sonic Intimacy*. New York: Bloomsbury Academic.

James, Robin. 2019. *The Sonic Episteme: Acoustic Resonance, Neoliberalism, and Biopolitics*. Durham: Duke University Press.

Jennings, Ros, and Abigail Gardner. 2012. *Rock On: Women, Ageing and Popular Music*. Farnham: Ashgate.

Jones, Reece. 2017. *Violent Borders. Refugees and the Right to Move*. London, New York: Verso.

Jones, Susanne. 2011. "Supportive Listening." *International Journal of Listening* 25 (1–2): 85–103.

Jorgensen, Martin Bak, and Carl-Ulrik Schierup. 2021. "Transversal Solidarities and the City: An Introduction to the Special Issue." *Critical Sociology* 47 (6): 845–55. https://doi.org/10.1177/0896920520983418.

Kanngieser, Anja. 2013. "Tracking and Tracing: Geographies of Logistical Governance and Labouring Bodies." *Environment and Planning D: Society and Space* 31 (4): 594–610.

Kanngieser, Anja. 2015. "Geopolitics and the Anthropocene: Five Propositions for Sound." *Geohumanities* 1 (1): 80–5.

Kaplan, Ann E. 1987. *Rocking Around the Clock Music Television, Postmodernism, and Consumer Culture*. New York: Routledge.

Kassabian, Anahid. 2013. *Ubiquitous Listening: Affect, Attention and Distributed Subjectivity*. Berkeley: University of California Press.

Keough, Sara Beth. 2010. "The Importance of Place in Community Radio Broadcasting: A Case Study of WDVX, Knoxville, Tennessee." *Journal of Cultural Geography* 27 (1): 77–98.

Kheshti, Roshanak. 2015. *Modernity's Ear: Listening to Race and Gender in World Music*. New York: New York University Press.

Kjeldsen, Svend. 2020. "Mancunian Irish: Identity, Cultural Intimacy, and Musican Hybridization – Urban Ethnomusicology and Cultural Mapping." In *Cultural Mapping and Musical Diversity*, edited by Britta Sweers and Ross Sarah, 35–53. Sheffield: Equinox.

Kristeva, Julia. 1982. *Powers of Horror: An Essay on Abjection*. New York: Columbia University Press.

Krüger, Simone and Ruxandra Trandafiou. [eds] 2014. *The Globalisation of Musics in Transit: Music, Migration and Tourism*. New York, London: Routledge.

Kuhn, Annette. 2002. *Family Secrets: Acts of Memory and the Imagination*. London, New York: Verso.

Kuhn, Annette. 2007. "Photography and Cultural Memory: A Methodological Approach." *Visual Studies* 22 (3): 283–92.

Kuhn, Annette. 2010. "Memory Texts and Memory Work: Performances of Memory in and with Visual Media." *Memory Studies* 3 (4): 298–313. https://doi.org/10.1177.

LaBelle, Brandon. 2006. *Background Noise: Perspectives on Sound Art*. New York, London: Continuum.

LaBelle, Brandon. 2010. *Acoustic Territories: Sound Culture and Everyday Life*. New York: Bloomsbury.

LaBelle, Brandon. 2018. *Sonic Agency: Sound and Emergent Forms of Resistance*. London: Goldsmiths Press.

Lambert, Joe. 2013. *Digital Storytelling - Capturing Lives Creating Community*. 4th ed. London: Routledge.

Leadsom, Andrea. 2022. *Twitter*. https://twitter.com/home.

Lee, Wonseok, and Grace Kao. 2021. "You Know You're Missing Out on Something: Collective Nostalgia and Community in Tim's Twitter Listening Party during COVID-19 Wonseok Lee &Grace Kao Pages 36–52 | Published Online: 08 Dec 2020." *Rock Music Studies* 8 (1): 36–52.

Lefebvre, Henri. 2004. *Rhythmanalysis: Space, Time and Everyday Life*. London: Bloomsbury.

Lloyd, Justine. 2009. "The Listening Cure." *Continuum* 23 (4): 477–87. https://doi.org/10.1080/10304310903003288.

Long, Emma, Annabelle Edwards, Brigit McWade, Samuel Clark, and Liz Brewster. 2021. "Older Veterans: The Materiality of Reminiscence, Making Unknown Histories Knowable and Forging Social Connections." *Memory Studies* 14 (4): 892–908.

Lundby, K., ed. 2008. *Digital Storytelling, Mediatized Stories: Self-representations in New Media*. New York, etc.: Peter Lang.

Lykes, M. Brinton, M. Emilia Bianco, and Gabriela Távara. 2020. "Contributions and Limitations of Diverse Qualitative Methods to Feminist Participatory and Action Research with Women in the Wake of Gross Violations of Human Rights." *Methods in Psychology* 4. https://doi.org/10.1016/j.metip.2020.100043.

Lykke, Nina. 2010. *Feminist Studies: A Guide to Intersectional Theory, Methodology and Writing*. New York: Routledge.

Macarthur, Sally. 2016. "Immanent listening." In *Music's Immanent Future: The Deleuzian Turn in Music Studies*, edited by S. Macarthur, J. I. Lochhead, and J. R. Shaw, 140–6. Oxford, New York: Routledge.

Macarthur, Sally, Judy Lochhead, and Jennifer Shaw, eds. 2016. *Music's Immanent Future: The Deleuzian Turn in Music Studies*. Oxford, New York: Routledge.

Margalit, Avishai. 2002. *The Ethics of Memory*. Cambridge, MA: Harvard University Press.

Marsilli-Vargas, Xochitl. 2014. "Listening Genres: The Emergence of Relevance Structures through the Reception of Sound." *Journal of Pragmatics* 69: 42–51.

Martin. 2019. "Lange Nacht in DeutschlandFunk." *Gundi Forum* (blog). ket.

Martin, Richard. 2021. "The Protest Provisions of the Police, Crime, Sentencing and Court Bill: A 'Modest Reset of the Scales'?" LSE Legal Studies Working Paper No. 15. https://doi.org/10.2139/ssrn.3973201.

Massumi, Brian. 1995. "The Autonomy of Affect." *Cultural Critique* 31: 83–109.

Mbembe, J. A., and Libby Meintjes. 2003. "Necropolitics." *Public Culture* 15 (1): 11–40.

Mendívil, Julio. 2017. "Schlager and Musical Conservatism in the Post-War Era." In *German Pop Music: A Companion - Companions to Contemporary German Culture*, edited by Uwe Schutte, 25–42. Berlin, Boston: De Gruyter.

Mendívil, Julio, Christophe Jacke, and M. Ahlers, eds. 2017. "Rocking Granny's Living Room? The New Voices of German Schlager." In *Perspectives on German Popular Music*, 100–8. New York, Oxford: Routledge.

Merleau-Ponty, Maurice. 1964. *The Primacy of Perception*. Northwestern University Press.

Moglen, Helen. 2008. "Ageing and Transageing: Transgenerational Hauntings of the Self." *Studies in Gender and Sexuality* 9: 297–311. https://doi.org/10.1080/15240650802370668.

Moreno, Jairo. 2016. "Imperial Aurality: Jazz, the Archive, and U.S Empire." In *Audible Empire: Music, Global Politics, Critique*, edited by Ronald Radano and Tejumola Olaniyan. 135–60. Durham: Duke University Press.

Moten, Fred. 2003. *In the Break: The Aesthetics of the Black Radical Tradition*. Minneapolis: University of Minnesota Press.

Murdock, Maureen. 2003. *Unreliable Truth. On Memoir and Memory*. New York: Seal Press.

Nancy, Jean-Luc. 2007. *Listening*. New York: Fordham University Press.

Needham, Gary. 2009. "Scheduling Normativity: Television, the Family, and Queer Temporality." In *Queer TV: Theories, Histories, Politics*, edited by Glyn David and Gary Needham, 143–58. Abingdon: Routledge.

Novak, David, and Matt Sakakeeny, eds. 2015. *Keywords in Sound*. Durham: Duke University Press.

Nugin, Raili. 2021. "Rejecting, Re-Shaping, Rearranging: Ways of Negotiating the Past in Family Narratives." *Memory Studies* 14 (2): 197–213.

O'Keefe, Linda, and Isabel Nogueira, eds. 2022. *The Body in Sound, Music and Performance: Studies in Audio and Sonic Arts*. New York, London: Routledge.

Oliver, Kelly. 2015. "Witnessing, Recognition, and Response Ethics." *Philosophy and Rhetoric* 48 (4): 473–93.

Osibisa. 1971. *Woyaya*. Vinyl. Germany: MCA.

Ouzounian, Gascia. 2021. *Stereophonica: Sound and Space in Science, Technology, and the Arts*. Cambridge, MA: MIT Press.

Oxford Department of International Development. n.d. "Aural Borders." Oxford. https://www.qeh.ox.ac.uk/content/aural-borders-audible-migrations-sound-and-citizenship-athens.

Paterson, Mark. 2007. *The Senses of Touch. Haptics, Affects and Technologies*. Oxford, New York: Bloomsbury.

Polychroniou, C. J. 2022. "Global Policy Journal." *We Need a World Without Borders on Our Increasingly Warming Planet* (blog). September 13, 2022. https://www.globalpolicyjournal.com/blog/13/09/2022/we-need-world-without-borders-our-increasingly-warming-planet.

Popa, Florian. n.d. *Galbene Gutuie*. Accessed July 14, 2022. https://www.youtube.com/watch?v=bjH_grBVeKU.

Poelzleitner, Elisabeth, Hermine Penz, and Roberta Maierhofer. 2019. 'Digital Storytelling in the Foreign Language Classroom'. In International Digital Storytelling Conference 2018 Proceedings. *Zakynthos: Club UNESCO Zakynthos*. https://dst.ntlab.gr/2018/proceedings

Protest Powers, Crime, Sentencing and Courts Act 2022 Factsheet. 2022. Available at https://www.gov.uk/government/publications/police-crime-sentencing-and-courts-bill-2021-factsheets/police-crime-sentencing-and-courts-bill-2021-protest-powers-factsheet.

Queen. 1978. *Spread Your Wings*. Vinyl. London: EMI.

Radano, Ronald, and Tejuomola Olaniyan, eds. 2016. *Audible Empire. Music, Global Politics, Critique*. Durham: Duke University Press.

Radstone, Susannah. 2007. *The Sexual Politics of Time: Confession, Nostalgia, Memory*. Oxford, New York: Routledge.

Radstone, Susannah, and Bill Schwarz. 2010. "Introduction: Mapping Memory." In *Memory: Histories, Theories, Debates*, edited by Susannah Radstone and Bill Schwartz. New York: Fordham University Press.

Rajchman, John. 2000. *The Deleuze Connections*. Cambridge, MA: The MIT Press.

Ramasamoglu, Caroline, and Janet Holland. 2002. *Feminist Methodology: Challenges and Choices*. London: Sage.

Ratcliffe, Krista. 1999. "Rhetorical Listening: A Trope for Interpretive Invention and a 'Code of Cross-Cultural Conduct.'" *College Composition and Communication* 51 (2): 195–224.

Ratcliffe, Krista. 2000. "Eavesdropping as Rhetorical Tactic: History, Whiteness, and Rhetoric." *JAC* 20 (1): 87–119.

Ratcliffe, Krista. 2005. *Rhetorical Listening: Identification, Gender, Whiteness*. Carbondale: Southern Illinois University Press.

Regev, Motti. 2013. *Pop Rock Music: Aesthetic Cosmopolitanism in Late Modernity*. Cambridge: Polity.

Rice, Tom. 2015. "Listening." In *Keywords in Sound*, edited by David Novak and Matt Sakakeeny. Durham: Duke University Press.

Ricoeur, Paul. 1984. *Time and Narrative*, vol. 1. Chicago: University of Chicago Press.

Robin, B. 2008. "Digital Storytelling: A Powerful Technology Tool for the 21st Century Classroom." *Theory Into Practice* 47: 220–8.

Robinson, Fiona. 2011. "Stop Talking and Listen: Discourse Ethics and Feminist Care Ethics in International Political Theory." *Milennium. Journal of International Studies* 39 (3): 845–60.

Roquet, Paul. 2021. "Acoustics of the One Person Space: Headphone Listening, Detachable Ambience, and the Binaural Prehistory of VR." *Sound Studies* 7 (1): 42–63. https://doi.org/10.1080/20551940.2020.1750270.

Rosenberg, Buck Clifford. 2016. "Shhh! Noisy Cities, Anti-Noise Groups and Neoliberal Citizenship." *Journal of Sociology* 52 (2): 190–203. https://doi.org/10.1177/1440783313507493.

Rothberg, Michael. 2009. *Multidirectional Memory, Remembering the Holocaust in the Age of Decolonization*. Stanford: Stanford University Press.

Schafer, Katharina, and Tuomas Eerola. 2020. "How Listening to Music and Engagement with Other Media Provide a Sense of Belonging: An Exploratory Study of Social Surrogacy." *Psychology of Music* 2 (48): 232–51.

Schulte, Christopher. 2016. "Possible Worlds: Deleuzian Ontology and the Project of Listening in Children's Drawing." *Cultural Studies Critical Methodologies* 16 (2): 141–50. https://doi.org/10.1177/1532708616636615.

Schulze, Holge, ed. 2020. *The Bloomsbury Handbook of the Anthropology of Sound*. London, New York: Bloomsbury Academic.

Schwarz, Bill. 2005. "'Already The Past': Memory and Historical Time." In *Memory Cultures: Memory, Subjectivity and Recognition*, edited by Susannah Radstone and Katharine Hodgkin, 135–51. Transaction Publishers.

Scrimshaw, W. 2013. "Non-cochlear Sound: On Affect and Exteriority." In *Sound, Music, Affect: Theorizing Sonic Experience*, edited by M. Thompson and I. Biddle, 27–43. New York: Bloomsbury Academic.

Sevenhuijsen, Selma. 2003. "The Place of Care: The Relevance of the Feminist Ethic of Care for Social Policy." *Feminist Theory* 4 (2): 179–97.

Sky News. 2022. "Noisy Protesters Could Feel Full Force of the Law after New Bill Is Approved," 27 April 2022. https://news.sky.com/story/noisy-protesters-could-feel-full-force-of-the-law-after-new-bill-is-approved-12599806.

Sloterdijk, Peter. 2016. *Weltfremdheit*. Frankfurt: Suhrkamp Verlag.

Solnit, Rebecca. 2020. "Respectfully." *Humans and Nature*. Accessed January 28, 2022. https://www.humansandnature.org/respectfully.

Sontag, Susan. 2003. *Regarding the Pain of Others*. London: Penguin.

Spivak, Gayatri. 1988. *Can the Subaltern Speak?* Basingstoke.

Stead, Michelle. 2016. "Seeing the Sense: Imagining a New Approach to Acousmatic Music and Listening." In *Music's Immanent Future: The Deleuzian Turn in Music Studies*, edited by Sally Macarthur, Judy Lochhead, and Jennifer Shaw, 179–87. London, New York: Routledge.

Spivak, Gayatri. 2010. "Can the Subaltern Speak?" In *Can the Subaltern Speak?: Reflections on the history of an idea*, edited by Rosalind. C. Morris. New York: Columbia University Press.

Sterne, Jonathan. 2003. *The Audible Past: Cultural Origins of Sound Reproduction*. Durham: Duke University Press.

Stewart, Kathleen. 2007. *Ordinary Affects*. Durham, London: Duke University Press.

Stewart, Kathleen. 2008. "Weak Theory in an Unfinished World." *Journal of Folklore Research* 45 (1): 71–82.

Stoever, Jennifer Lyn. 2016. *The Sonic Colour Line: Race and the Cultural Politics of Listening*. New York: New York University Press.

Stratton, Jon, and Nabeel Zuberi. 2014. *Black Popular Music in Britain Since 1945*. Ashgate Popular and Folk Music Series. Farnham: Ashgate.

Suso, Foday Musa. 1998. *Jooka Tamala*. Vinyl. PopLlama Records.

Sweers, Britta, and Sarah Ross. 2020. *Cultural Mapping and Musical Diversity*. Sheffield: Equinox Publishing Limited.

Szendy, Peter. 2008. *Listen. A History of Our Ears*. New York: Fordham University Press.

Szendy, Peter. 2017. *All Ears. The Aesthetics of Espionage*. Fordham University Press.

Tandeciarz, Silvia R. 2006. "Mnemonic Hauntings: Photography as Art of the Missing." *Social Justice* 33 (2): 135–52.

Tanner, Grafton. 2021. *The Hours Have Lost Their Clock*. London: Repeater.

Taylor, Diana. 2003. *The Archive and the Repertoire: Performing Cultural Memory in the Americas*. Durham, London: Duke University Press.

Taylor, Harry. 2022. "Dominoes Player Wins Case over 'Racist' Noise Ban in London Square." *The Guardian*, May 14, 2022. https://www.theguardian.com/uk-news/2022/may/14/dominoes-player-wins-case-over-racist-noise-ban-in-london-square.

Taylor, Timothy D. 2005. "Radio in Twenties America: Technological Imperialism, Socialization, and the Transformation of Intimacy." In *Wired for Sound: Engineering and Technologies in Sonic Cultures*. Middletown: Weslyan University Press.

The Home Office. 2022. "Noise-Related Provisions: Police, Crime, Sentencing and Courts Act 2022 Factsheet." Home Office. https://www.gov.uk/government/publications/police-crime-sentencing-and-courts-bill-2021-factsheets/police-crime-sentencing-and-courts-bill-2021-noise-related-provisions-factsheet.

"The Listening Project." 2012a. https://www.bbc.co.uk/programmes/articles/41rDvmTW0T1JWjXkcvZtMqt/about.

"The Listening Project." 2012b. https://sounds.bl.uk/Oral-history/The-Listening-Project.

Thompson, Marie. 2017. *Beyond Unwanted Sound: Noise, Affect and Aesthetic Moralism*. London: Bloomsbury Academic.

Thornam, Sue. 2001. *Feminist Theory and Cultural Studies: Stories of Unsettled Relations*. London: Arnold.

Tobi, Thomas. 2022. "West Indian Dominoes Players Dismayed by Noise Ban in London Square." *The Guardian*, May 7, 2022. https://www.theguardian.com/uk-news/2022/may/07/west-indian-dominoes-players-dismayed-noise-ban-london-square.

Toltz, Joseph. 2016. "Listening to Ethnographic Holocaust Musical Testimony through the Ears." In *Music Immanent Futures: The Deleuzian Turn in Music Studies*, edited by Sally Macarthur, Judy Lochhead, and Jennifer Shaw, 188–200. Routledge.

Tuuri, Kai, and Thomas Eerola. 2012. "Formulating a Revised Taxonomy for Modes of Listening." *Journal of New Music Research* 41 (2): 137–52. https://doi.org/10.1080/09298215.2011.614951.

Tyler, Bonnie. 1984. *Holding Out for a Hero*. Europe: CBS.

UK Home Office. 2021. "Police, Crime, Sentencing and Courts Bill 2021: Protest Powers Factsheet." London. https://www.gov.uk/government/publications/police-crime-sentencing-and-courts-bill-2021-factsheets/police-crime-sentencing-and-courts-bill-2021-protest-powers-factsheet.

UK Parliament. n.d. *Police, Crime, Sentencing and Courts Bill Amendments to the Public Order Bill*. Accessed August 27, 2022. https://bills.parliament.uk/bills/2839.

van Dijck, José. 2014. "Remembering Songs through Telling Stories: Pop Music as a Resource for Memory." In *Sound Souvenirs: Audio Technologies, Memory and Cultural Practices*, edited by Karin Bijsterveld and José van Dijck, 107–20. Amsterdam: Amsterdam University Press.

Voegelin, Salome. 2010. *Listening to Noise and Silence: Towards a Philosophy of Sound Art*. New York, London: Bloomsbury Academic.

Voegelin, Salome. 2014. *Sonic Possible Worlds*. New York, London: Bloomsbury.

Voegelin, Salome. 2019. *The Political Possibility of Sound: Fragments of Listening*. New York, London: Bloomsbury Academic.

Walia, Harsha. 2021. *Border and Rule Global Migration, Capitalism, and the Rise of Racist Nationalism*. Chicago: Haymarket Books.

Walkerdine, Valerie. 1997. *Daddy's Girl: Young Girls and Popular Culture*. Cambridge, MA: Harvard University Press.

Walkerdine, Valerie, Helen Lucey, and June Melody. 2002. "Subjectivity and Qualitative Method." In *Qualitative Research in Action*, edited by T. May, 179–99. London: Sage.

Walmsley, Andy. 2011. "Random Radio Jottings." http://andywalmsley.blogspot.com/2011/08/are-you-sitting-comfortably.html.

Weheliye, Alexander G. 2005. *Phonographies: Grooves in Sonic Afro-Modernity*. Durham: Duke University Press.

Welzer, Harald. 2010. "Re-Narrations: How Pasts Change in Conversational Remembering." *Memory Studies* 3 (1): 5–17.

Western, Tom. 2020. "Listening with Displacement: Sound, Citizenship, and Disruptive Representations of Migration." *Migration and Society: Advances in Research* 3: 294–309.

whirligig tv co. 2011. "Listen With Mother." *Listen With Mother (1950–82)*. http://www.whirligig-tv.co.uk/radio/lwm.htm.

Yuval-Davis, Nira. 1999. "What Is 'Transversal Politics'?" *Soundings* 12: 94–8.

Index

Abject 118
acousteme 24
acoustic resonance 25, 26
Aden Veterans Association 66, 71
Adorno, Theodor 15, 116
aesthetic cosmopolitanism 134
affect 31, 47, 82, 117, 126
affective atmospheres 118
affective connections 11, 40
affective contagion 47
affective economy 12, 46, 81, 89
Afro-Caribbean communities 27
Ageing Communication Technologies (ACT) 49
ageist stereotypes 70
Agency of Cultural Affairs 147
Age UK Gloucestershire 43, 58, 60, 61, 63–5, 67, 71, 75, 80
aging 2–3, 11, 43, 44, 54
Aging Activisms 49
Ahmed, Sara 4, 46, 81
Airstream caravan 98
Alfred, Willy 66, 70–3, 79, 80, 152
ambient citizenship 92, 111, 118
American hip hop 20
"amongst-ness" 30
Anders, Günter 24
Anger, Moses 82
Anthropocene 31
anti-Brexit marches 115
anti-nuclear marches (1980s) 58
Appadurai, Arjun 99
apps 131–2
Arendt, Hannah 40, 53, 56
Armed Forces Covenant Fund Trust 43, 63
articulated voice 146
artifacts 48, 50, 64, 65, 67, 73, 78
Art of Listening, The (Back) 3, 29
ArtSpace 110
asylum seekers 100, 121, 125, 152
Atkinson, Helen 65, 75

Attali, Jacques 26, 114
Audible Empire (Radano and Olaniyan) 27
audile techniques 22
audiopolitics 26, 27
audio stimuli 22
audiovisual listening 21
Auge, Marc 133
Auschwitz-Birkenau State Museum 83
Auschwitz Museum 33, 36, 40–2, 59, 63, 85, 87, 89
@AuschwitzMuseum 11, 12, 33, 36, 42, 63, 64, 81–8, 106
Austin, J. J. 35
authenticity 123
autobiography 57, 87
autoethnography 10

Back, Les 3, 29
Barad, Karan 18, 30, 33, 76
baroque recital 20
Barthes, Roland 4, 22
Barwick, Nick 54
Battle for Aleppo (2012–16) 143
Battle of Imjin 75, 79
Bauman, Zygmunt 77, 123
BBC archives 40, 96, 105
BBC Light Programme 93
BBC news 94
BBC Radio 91–3, 98, 106, 118
BBC Radio 4 93, 98, 106, 125
BedTime Stories (Moulton) 94
bedtime story 94
Befindlichkeit 24
being "among" 30
"being with" 30
belonging 39, 151
  borders of 92
  and listening 4–5
  and memory 10
Bergson, Henri 18, 33
Berlant, Lauren 92, 96, 111, 118

Beyoncé   26
Bickford, Susan   56, 147
Birdsall, Carolyn   18
Black British reggae systems   25, 26
Black masculinity   28
Blackness   28, 116, 117
Black youth   114
Blick, Chris   66, 67, 70, 74, 76, 78, 79, 152
#BLM event   20
body personal   28
body politic   28, 93
borders   131–2
    abolition of   112
    violence at   117
Born, Georgina   20, 152
Bosnian War (1992–5)   61, 64, 67, 76, 78
boundary collapse   24
Bourdieu, Pierre   47
Bovens, Mark   105
Bradley, Gracie Mae   112, 131
Braidotti, Rosi   30, 125
Bray, Steve   112, 113
Brexit   104
British Academy   45, 60, 61
British Armed Forces   65
British Library   40, 92, 96, 98, 100, 105, 106
British National Service   66, 72, 74
Britishness   106
British war in Aden (1963–7)   64, 66, 68, 69, 76–9
Bulgaria   12, 124, 128
Bull, Michael   6, 30
Butler, Judith   6, 35

"calamity"   27
Caminos   124
Campt, Tina M.   11, 18, 36, 41, 51, 85
Canetti, Elias   18
capitalism   22
caregiving   58–61
carnivals   27
Carter, Paul   42
"Ça va" (song)   142
Cesaro, Steph   23
Charmaz, Kathy   55
Chazan, May   49

Cheltenham Hebrew Congregation   1, 57, 153
Cheltenham Pump Rooms   19
Cheltenham Spa's Regency   19–20
Chion, Michel   21
Choukeir, Samih   143, 144
Chouliaraki, Lilie   112, 137, 145, 146
civvy street   65, 79
climate change   45, 114
Clough, Tommy   73–5, 77–80
Coaker, Lord Vernon Rodney   113
Cohen, Sara   69, 125
colonialism   25, 117
colonial subjectivity   145
Concordia University   49
connected listening   7, 10–11, 39–62, 88, 93, 105, 115, 118, 125, 148
    digital storytelling and   47–50
    as double witness   53–4
    Grounded Feminist Listening   54–8
    and image   50–3
    listening and discomfort   43–7
    listening as care   58–61
Conservative Party   29, 113, 114
contemporary British citizenship   91, 116
conversational remembering   74
Cornwall   43, 45, 58
Couldry, Nick   17
Countryside and Community Research Institute (CCRI)   45, 107
Covid-19 pandemic   40, 57, 64, 65, 71, 128
Creed, Barbara   117
Criminal Justice Bill (1994)   114
cross-cultural communication   60, 73
CSC Danilo Dolci   124
CSI   124
cultural cafés   147
culture(s)   58
    of belonging   91
    citizen   91
    mapping   132
    national   95
    philosophy   16
    popular   4
    process   117
    studies   4, 46, 59
Cumbria   43, 45, 50, 58

Cypriot immigration   127
Cyprus   12, 68, 124, 127, 128, 130, 133, 141, 142, 147

Davies, Eirlys E.   32, 42
defamiliarization   93
Deleuze, Gilles   15, 23, 42, 47
democracy   56, 105
Denning, Michael   26
De Nora, Tia   73, 136
De Noronha, Luke   112, 131
Derrida, Jacques   53
Diabate, Sidiki   141
Diabate, Toumani   141, 142
Digital Media and Film Production courses   65, 70
digital stories   63, 65, 66, 71, 107–9
Digital Stories   89
digital storytelling   1, 5, 8, 11, 17, 22, 36, 40–5, 47–50, 57, 60, 63, 65–8, 77, 78, 91, 93, 107, 110, 111, 136, 153
digital technologies   20
Diken, Bülent   145
discomforting spaces   27
discourse analysis   46
domesticity   95, 96
domestic labor   95, 96
domestic politics   58
Donovan, Kate   31
double witnessing   11, 53–4
Dowdeswell, Alan   66–8, 72–4, 78, 152
Dreher, Tanja   41, 54
Dunford, Mark   49
Dyer, Richard   57

editing   47–9, 57, 65, 66, 68, 74, 78, 80
Eerola, Thomas   21
Eidsheim, Nina Sun   32
embodiment   31
emotions   81, 126
empathetic listening   20, 31, 40
"encoding, decoding" (Hall)   8
episteme   24–6
Erasmus+   28, 44, 50, 66, 108, 111, 122, 123, 125, 126, 148
Erlmann, Veit   23–5, 28
Ettore, Elizabeth   10
EU Commission Reports   127

EU Indicators of Immigrant Integration   127
EU Referendum (2016)   115
European Court of Human Rights   121
European Social Fund   107
European Union   148
expanded/expansive listening   31, 32
experimental radio   31
Extinction Rebellion   114

Fairuz   140
familial intimacies   93
Feldman, Suzanne   82, 89, 152
Felman, Shoshana   5, 86
felt materiality   137
feminism/feminist   54, 55
   autoethnography   55
   care ethics   11, 59, 151
   cultural studies   54, 55
   intergenerational   55
"feminist 'I'"   10
Ferrandiz, Francisco   33
Ferrara, Beatrice   27
Finer, Ella   16, 18
First World War (1914–19)   78
Fivush, Robyn   73, 105
Floyd, George   20
Forman, Murray   118
Foucault, Michel   18
Fraser, Robert   1, 123
Fuss, Diane   42

"Galbene Gutuie" (song)   138–9
Gallagher, Michael   23, 31, 114
GARAS   107
Garcia, Luhushis-Manuel   126
Garde-Hansen, Joanne   1, 59, 97
Gateway Trust   44
Gaynor, Gloria   142
gender   5, 27, 35, 54, 59, 95
George III (King)   20
Georgiou, Myria   112, 137, 145, 146
Gloucester Deaf Association   107
Gloucester Farmers' Club   65
Gloucestershire Aden Veterans Organization   79
Gloucestershire Asylum   107
Gloucestershire County Council   44

Glover, Fi 98, 99
Going the Extra Mile (GEM) 44, 51, 52, 58, 60, 61, 107–11, 119
good citizen 12, 93, 110, 115
Google Map 61, 130
Gorton, Kristyn 59, 97
The Government's Integration Goals 127
Greece 12, 44, 53, 124, 128, 144, 147
Greenpeace 114
Grenier, Line 69, 125
Grosz, Elizabeth 25, 117
Grounded Feminist Listening 11, 42, 54–8
Guattari, Felix 15, 23, 42, 47
gwerz 141

Halberstam, Judith 97
Halbwachs, Maurice 86
Hall, Stuart 8, 42
Hanachi, Abdelwahab 139
Hansen, Kai Arne 124, 127, 143
"haptic image" 36
Haraway, Donna 56
Harrow Road 115
Hatcher, Faye 99, 105
Hayball, Julia 50, 66
Health and Social Care, University of Gloucestershire 64, 65
Hebdige, Dick 115
Henriques, Julien 27
hermeticism 24
heteronormative weddings 60
heteronormativity 4, 5, 96
"An Hini A Garan" (song) 140–2
Hirsch, Marianne 18
historiography 35
"Holding Out for a Hero" (song) 140
Holocaust 5, 11, 12, 18, 36, 46, 54, 63, 81, 85, 86, 88, 89, 152
1933–45 63
Home Office 113
hooks, bell 5
house parties 27
How to Do Things with Words (Austin) 35
Hristova, Marije 18, 33
Hrycak, Alexandra 17
Hutt, E. R. 95

Illouz, Eva 47
impressions 146–8
"Inheritance Tracks" 125, 126
Inland Norway University of Applied Sciences 124
integration policy 127, 128
intergenerational conversations 11, 43, 49, 64
intersubjectivity 23
Irigaray, Luce 30
ISIS 79
Islam, Azharul 57, 78
Istvandity, Laura 86, 133
Italy 12, 44, 50, 124, 128
"items of memorabilia" 65
"I Will Survive" (song) 142–3

James, Martin 25
Jarrow March (1936) 115
Jennings, Ros 69, 125
Jews 81, 82, 84
Joe Lambert's Story Center 48
Joining Forces 43, 63–5
Jones, Reece 117
Jones, Susanne 92

Kanngieser, Anja 31
Kapchan, Deborah 27
Kassabian, Anahid 8, 22, 23, 30, 36, 37
"keep it short and simple" (KISS) model 50
Khan, Sadiq 113
Kheshti, Roshanak 27, 59
Kigali 121
kinaesthetic listening 31
KMOP 124, 133
Know and Can 124, 128
Korean Prisoner of War camp 78
Korean War (1950–3) 64, 66, 68, 74, 75, 78
Kristeva, Julia 117, 118
Kuhn, Annette 8, 18, 34, 70, 72, 74, 77, 85
Kum, Talita 142
Kvall, Camilla 124

LaBelle, Brandon 7, 27, 41, 47, 59, 123, 136, 146, 147, 149

Labor Party   43, 58
"Lampedusa" (song)   141, 142
Landberger, Robert   1–3, 13, 152, 153
Lang, Julia   93
"The Last Post"   69, 77
Laub, Dori   5, 86
LBC radio   113
Leadsom, Andrea   113, 114
Lefebvre, Henri   33, 34
Levy, Genevieve Annette Flora   82, 89, 152
listening   15, 16. *See also specific entries*
  across time/gender/class   25–9
  acts of   17, 39
  affective   111, 132
  with age   44
  and attention   40
  cognitive and emotional   47
  concept(s) of   16, 23
  as connecting process   21
  convivial   52, 111, 119, 149
  corporate   42
  as cross-disciplinary methodology   3
  dialogic   42
  emergent   32, 47
  ethical   32
  fluidity and connectivity   30
  generative   32
  vs. hearing   22, 23
  to images   11, 35–6
  in-between   30–1
  interfaces and interactions   9–10
  layers of   28
  and lines of familiarity   132–4, 136
  maternal   21, 22
  and memory   5–6
  as methodology of care   31
  multimodal   23
  narrative and memory   34–5
  online   11, 87
  orders   22–3
  to past   32–4
  performative   18
  as practice   29, 33, 36
  process   97
  and proximity   6, 21, 27, 30, 33, 149
  relational   32
  relationship to embodied self   17
  relationship with past   18
  rhetorical   41
  seeing and   24
  and story   8
  types   18–21
  to voices   23
listening across age(s)   63–89
  @AuschwitzMuseum   81–8
  listening with age   80–1
  performative historiography   74–6
  remembering in place   72–3
  rippling   76–8
  upsetting narratives   78–80
  Veterans' Voices   64–72
listening and belonging   4–5, 12, 91–119
  employed listening   107–11
  Listening Project, The (2012–22)   98–107
  Listen with Mother (1950–82)   91–8
  not listening   112–18
"Listening Cure" (Lloyd)   42
Listening Project, The (2012–22, radio program)   12, 47, 91, 92, 98–107, 119
Listening to Images (Campt)   36
Listen with Mother (1950–82, radio program)   12, 91–8, 119
Lloyd, Justine   42
Long, Emma   72–4
Lykke, Nina   42, 56

Macarthur, Sally   19
McNab, Maddy   49
"Mahboubi" (song)   139
Maida Hill market square   115, 117
Maida Vale   115
MaMuMi app   127, 128, 130–2, 137, 144, 146
MaMuMi Migration Map   147
MaMuMi work packages   126–31
Mapping the Music of Migration (MaMuMi) Song Story workshop   12, 26, 28, 34, 44, 50, 52, 57, 60, 61, 122–6, 128–30, 133, 144–8, 152
maps/mapping   131–2
Margalit, Avishai   85
maternal/child intimacy   97–8

maternal identity 96
media messages 8, 9
memoirs 78, 87
memories 39, 75, 78, 88, 115, 151
    affective 132
    and belonging 10
    collective 86
    and listening 5–6
    musical 81, 133
    national 72
    shared 85
    social 74
    studies 5, 10, 16–18, 37, 76
    time and 33
"Memory Worlds: Reframing Time and the Past" (Hristova, Ferrandiz, and Vollmeyer) 33
Mercury, Freddie 138
Merleau-Ponty, Maurice 55
micro publics/micropolitics 17
migrants 12, 19, 26, 39, 52, 53, 56, 57, 61, 62, 104, 122–5, 127, 128, 145, 152
migration 25, 46, 92, 100–4, 112, 117, 121–4, 127
Miners protest (1984) 115
Ministry of Children, Equality and Social Inclusion 127
mobile Listening Booth 98
Moglen, Deborah 2, 33
Moorey, Gerry 99
"moral geography of noise" 115
Moten, Fred 36
Mozes, Léon 85, 152
Murdock, Maureen 75
music
    events 127–8
    experience 152
    genre 134–5
    listening 20, 21, 59
    practices 3
    revolution 26
musicality 27
Musk, Elon 12
muting 18
"My Story" project 50, 108

Nancy, Jean-Luc 26, 28, 29
narratable selves 148

narratives 8, 16, 18, 31, 34–6, 40, 46, 49–53, 64, 74, 77, 78–80, 89, 105, 118, 123, 130, 134
National Age UK 65
national identity 95
Navigator Developers 107
Needham, Gary 97
neoliberal biopolitics 25, 26
neo-liberal capitalist model 110
neo-liberal economy 111
NetZero 45, 50, 60
NetZero British Academy 45
noise 27, 92, 93, 112–17, 121
noisesome-ness 27
noisy protest 6, 12, 112–14
non-citizens 92
non-governmental organization (NGO) 49, 52, 57, 61, 124, 125, 128–30, 146–8
non-hierarchical co-production model 48
Norway 12, 124, 128, 130, 133, 147
"not listening" 8–9, 92, 112–18
Notting Hill Carnival 27

offshore refugees 121
Olaniyan, Tejuomola 27, 117
oral history 3, 78, 99
ordinariness 82, 92, 106, 107, 122, 151
Ordinary Affects (Stewart) 31
Orgreave battles (1984) 115
"Other" music 59
Otherness 6, 10, 13, 18, 25, 27, 28, 116, 122
Oxenford, Daphne 93

Paunescu, Adrian 138, 139
Pennah, Emlyn 73, 78–80
performative historiography 8, 11, 18, 35, 64, 73–6, 89, 151
personal histories 74
photographic audibility 51
photographs 1, 12, 18, 35, 36, 41, 48, 50, 51, 63, 65, 71–3, 78, 81–3, 86, 108
Pittville Pump Room 20
Police, Crime, Sentencing and Courts Bill (2022) 29, 112
"politics of location" 56

"politics of recognition" 54
Poll Tax Riots (1991) 115
pollution 117
Pop, Florian 138
popular music 4, 5, 10, 11, 17, 18, 25, 46, 55, 134
power dynamics 49, 57
Premiere Pro 48, 70, 110
Prior, Jonathan 31
profound transformations 20
psychoanalytic framework 54

Qualitative Online Listening 11, 42, 46, 63, 88
Queen 137, 138

race 27, 117
racism 22
Radano, Ronald 27, 117
Radio Gloucestershire 106
The Radio Times 94, 97
Ratcliffe, Krista 41, 56, 60
reason and resonance 24, 25
recording 35, 40, 45, 46, 48–50, 57, 60, 65–9, 78, 128, 133
Refugee Association 107
refugees 57, 61, 100–4, 121, 124, 125, 145
Regarding the Pain of Others (Sontag) 35
Reith, John 91
"relocation of politics" 105
residential ethics 115
Responsibility of Forms, The (Barthes) 4
Rewakowicz, Maria G. 17
Rhythmanalysis (Lefebvre) 33
Rihanna 26
rippling 11, 64, 76–8
Rosenberg, Buck Clifford 115
Ross, Sarah 132
Rovelli, Carlo 33
Rutter, Rick 66, 70, 71, 74, 75, 78, 152
Rwanda 121, 149

Sajovics, Rachel (Geist) 84
Sawchuk, Kim 49
Sawicki, Pawel 83, 84, 88
School of Health and Social Care 43
School of Media, University of Gloucestershire 43, 64, 70

Schulte, Christopher 47
Schwartz, Bill 18, 33
scripting 34, 35, 48, 49, 65–9, 74, 108
Second World War (1939-45) 78, 110
self 34, 35, 37, 54, 58, 87, 105
Sevenhuijsen, Selma 105
Shaeffer, Pierre 21
silence 121, 123, 151
"situated knowledges" 56
Sloterdijk, Peter 7, 33
Smith, Dorothy 93, 95
social media 83, 84, 86, 87, 89, 114, 151
Soldiers, Sailors, Airmen and Families Association 64
Soldiers Museum Gloucester 50, 57, 65
Solnit, Rebecca 39, 41, 61
song stories 130–3, 136–46, 149
sonic agency 12, 25, 27, 40, 41, 111
sonic control 18
sonic soundscapes 26, 27
Sontag, Susan 35, 36, 85
sound 123, 126
   fluidity and flow 27–8
   methexic 29
   racializing 117
   and relationship with politics of race 25
   studies 10, 15–18
"spacetimemattering" 33
Spain 12, 103, 124, 127
Spitzer, Leo 18
Spivak, Gayatri 145
"Spread Your Wings" (song) 137–8
SSAFA 43
SSFA 64, 71
Stead, Michelle 8, 20, 23
Sterne, Jonathan 22
Stewart, Kathleen 31, 34, 46, 47, 81, 82, 88, 126, 134, 147, 149
"Stop Brexit Man" 112
stories 63, 64, 74–6, 78–81, 95, 124, 133
storytelling 3, 125, 132, 148
strategic idealism 41
stripping 112
subaltern sonic agencies 26
sudden eruptions 126
'The Sunday School Outing' (1958) 95
Suso, Foday Musa 137
Sweers, Britta 132

sympathetic listening   20
Syrian war   143

"Tamala" (song)   137
Tandeciarz, Silvia R.   85–6
Tannen, Deborah   60
Tanner, Grafton   73
Taylor, Timothy   94
Theophile, Ernest   115–18
"theory of moments"   33
Theresienstadt   63
Thompson, Marie   31, 116
Toltz, Joseph   28, 81
Tracichleb, Antoni   82
trans-aging   2, 33
trauma   5, 36, 54, 58, 64, 142, 144, 145
Troubling Inheritances (Cohen, Grenier, and Jennings)   125, 133
Tsolaki, Annita   143, 148
Tuuri, Kai   21
Twitter   6, 12, 36, 40–2, 46, 59, 63, 81–5, 87, 89, 106, 113, 152
Tyler, Bonnie   140

ubiquitous listening   20
UK   121, 124, 128, 137, 148, 152, 153
UK government   29, 91
UK House of Commons   113
UK House of Lords   29, 113
UK National Lottery Community Fund   61, 107
undesired sound politics   25
UNESCO   132
University Funding Office   44
University of California   48
University of Gloucestershire   66, 107, 123–5, 130
unwanted sound   116

VE Day   75, 78, 79
ventriloquized voice   146
Veterans' Voices   11, 33–5, 43, 44, 47, 48, 50, 52, 59, 60, 63–74, 76, 88, 89
violence   117, 144
visual culture   18
visual presence   29
Voegelin, Salome   22, 30, 32, 57, 137
voice(s)   31, 40, 121–3, 144, 145, 151
  Black   36
  mapping   144
  migrant   146
Volatile Bodies (Grosz)   25
Vollmeyer, Johanna   33

Wacholder, Dasza   87
Wacholder, Yoseph   87, 89
"Wakef Ya Askar Lubnan" (song)   140
Weltfremdheit (*Alienation*, Sloterdijk)   7
Welzel, Sally   68–9, 71, 74–9, 152
Welzer, Harald   74, 84
Westbourne Park   115
Western, Tom   117, 118, 145
Western classical traditions   20
Western philosophy   24
WhatsApp   133
whiteness   25, 57, 123
white patriarchy   25
Williams, Gwyneth   98
Windrush generation   116
Women Ageing and Media summer schools   125
Women's Royal Air Force (WRAF)   68
Woolaston, Brian   67, 74, 75, 79, 80
world music   27
www.mamumi.eu   137, 144

"Ya Hef" (song)   143

www.ingramcontent.com/pod-product-compliance
Lightning Source LLC
Chambersburg PA
CBHW052046300426
44117CB00012B/1999